ONE LONG EXPERIMENT

T0176418

Perspectives in Paleobiology and Earth History

The concept for these series was suggested by Mark and Dianna McMenamin, whose book *The Emergence of Animals* was the first to be published.

One Long Experiment

SCALE AND PROCESS IN EARTH HISTORY

RONALD E. MARTIN

COLUMBIA UNIVERSITY PRESS

NEW YORK

Columbia University Press
Publishers Since 1893
New York   Chichester, West Sussex
Copyright © 1998 Columbia University Press
All rights reserved

Library of Congress Cataloging-in-Publication Data

Martin, Ronald E.
    One long experiment : scale and process in earth history / Ronald E. Martin.
        p.   cm. — (Perspectives in paleobiology and earth history series)
    Includes bibliographical references (p.   –   ) and index.
    ISBN 0-231-10904-0 (acid-free paper). — ISBN 0-231-10905-9 (pbk. : acid-free
paper)
    1. Historical geology.   I. Title.   II. Series.
    QE28.3.M37   1997
    551.7—dc21   97-27821

Printed in the United States of America

c 10 9 8 7 6 5 4 3 2 1
p 10 9 8 7 6 5 4 3 2 1

To Carol and Dana,
with much love and deep appreciation for their tolerance

CONTENTS

CONTENTS

CONTENTS

PREFACE

This book is about many things. First and foremost, it is about history, especially in terms of process. In all its permutations, geology has always been the science of the history of the earth. Geologists, including stratigraphers and paleontologists, are historians. But why is the earth's history important? Its stratigraphic (fossil) record is basically one long (and poorly controlled) experiment that has followed a path contingent on its previous history. The record of those processes is the true value of the stratigraphic record because it gives us clues as to how we arrived at where we are today. Moreover, we must study not only the fossils and the phenomena themselves but also the limits to study of the processes recorded in the rocks.

Second, this book is about scale and hierarchy, and how geological processes that are detected vary with the scale used to measure them. The scales of geologic time and space offer perspectives not appreciated over a human lifespan. Geologic processes are not necessarily detectable over the timescale of generations, much less the Quaternary, but they are no less important, and perhaps more so, than the processes we do observe.

Third, the book is about the methodology of historical sciences such as paleontology and stratigraphy, and why that methodology is just as important as the reductionist approach to which most scientists—including historical scientists—have been indoctrinated.

In writing this book, I have tried to bear the following questions in mind:

- Based on the stratigraphic and paleontologic record, what is the evidence for processes that occur over spatio-temporal scales exceeding those of one or a few human generations?
- Why are these processes important to the human race?
- How does one test hypotheses using the stratigraphic (fossil) record?
- Given our present knowledge, what are the limits of that record?

Today study of the earth's history is under attack from many quarters, including some within the geological community who deem it to be irrelevant to environmental science. Much of geological history, perhaps with the exception of the last few million years, is now often viewed with the same ignorance and contempt accorded to the history of human civilization. Corporate and university administrators, who increasingly consider only the bottom line and who often confuse geology with geography because both bear the same prefix, have all too easily taken up this banner. Unfortunately, Earth historians have not done a very good job of communicating the importance of their profession to those who control the purse strings, and the relevance (an infamous buzzword of the 1960s) of their work is being questioned as a result.

Much of what I say is review, and much of it is not, but I hope that I have said it differently so that readers view this planet's history from a fresh perspective that further stimulates the historical science of geology. This book is intended as a primer and progress report, and I make no claims about its completeness. It could accompany a standard text in Earth science or it could stand alone in a seminar for advanced undergraduates and graduate students. The text is heavily referenced, especially chapters 7 and 8, both of which are often speculative (including some of my views), so that the reader can get into the primary literature quickly. I also discuss fractals, chaos theory, power laws, and self-organized criticality where appropriate, because I find these subjects fascinating and heuristically useful. However, I deemphasize jargon, in keeping with the conceptual, process-oriented approach of the book, although terms that are introduced are usually explained briefly. In chapter 3 I derive equations to give the reader a better chance of understanding the system under discussion (bioturbation), even at the risk of sounding pedantic. In some cases I have changed the standard symbols for certain variables or parameters so as to avoid confusion in the text. Many of the examples stem from my experience as a micropaleontologist, stratigrapher, and taphonomist, but I hope that students of other disciplines will find much to consider.

Any book as broad as this one necessarily suffers from errors of omission and interpretation, but my primary intent is to emphasize what I believe are important themes in Earth history and their (often neglected) interconnections. The book reflects the evolution of my approach to teaching paleontology and Earth history over the last decade, which has been described by one student as theoretical historical geology; for this reason I maintain a conversational tone throughout much of the book. Wherever possible I use figures instead of words, but for space and financial reasons in some cases I have instead cited figures within certain sources readily available to the reader.

I thank the National Science Foundation for its support of my research, which has led to this volume; Niles Eldredge and David Bottjer for their encouraging reviews of early versions of the manuscript; and Ed Lugenbeel, executive editor, and Alissa Bader for their support. Scott Hippensteel read through a late draft of the manuscript and found a number of errors; thanks also to Scott for his encouraging words and our many conversations about the Phillies and Reds.

I am also grateful to John Wehmiller for his support when things got tight; to Cheryl Doherty, of the University of Delaware Geology Department; and especially to Barbara Broge and Molly Chappel, of University of Delaware Media Design, for their heroic efforts with the figures. Carol Anne Peschke and Ron Harris made many helpful suggestions that improved the manuscript. Without all of them this book would not have come to fruition.

Ron Martin
Newark, Delaware

There are monstrous changes taking place in the world, forces shaping a future whose face we do not know. . . . In our time mass . . . production has entered our economics, our politics, and even our religion. . . . There is great tension in the world, tension toward a breaking point, and men are unhappy and confused. . . .

And now the forces marshaled around the concept of the group have declared a war of extermination on that preciousness, the mind of man. By disparagement, . . . by repressions, forced direction, and the stunning hammerblows of conditioning, the free, roving mind is being pursued, roped, blunted, . . .

I can understand why a system built on a pattern must try to destroy the free mind, for that is one thing which can by inspection destroy such a system. . . . I hate it and I will fight against it.
—*John Steinbeck,* East of Eden

Methodology and Proof in Historical Science

Newtonian tidiness . . . is coveted. —Allen and Starr

History is science of a different kind—pursued, when done well, with all the power and rigor ascribed to more traditional styles of science. —S. J. Gould

Historical sciences occupy the nebulous realm of the middle ground. They lie somewhere between small-number systems and large-number systems. The former treat their objects (such as planets) with the precision of differential equations (as Newton did), whereas the latter have so many objects to count (atoms, molecules) that only a statistical average is practical (gas laws). In either case, the basic approach is reductionist or mechanistic. If we merely simplify the system to its parts, we can understand the whole (Occam's razor or the principle of parsimony). In the case of historical sciences, there are too few parts to average their behavior and too many to account for separately with their own differential equation (Allen and Starr, 1982).

A common criticism of any historical science is that historians have a very difficult time proving anything. And if history cannot prove something, then it must not be science. "Good" science (meaning "hard" science, such as chemistry or physics), on the other hand, proceeds by "*the* scientific method," commonly but erroneously attributed to the highly influential twentieth-century philosopher Karl Popper (see Bondi, 1994): (1) observation, (2) hypothesis or tentative explanation (more than one hypothesis is preferable, but scientists, being human, favor their pet hypotheses), (3) experiment (to test the validity of the hypothesis), (4) modification or even rejection of the hypothesis, and (5) further testing of the new or modified hypothesis, all done under (preferably) highly controlled conditions (a laboratory is desirable, inhabited by a nearsighted scientist with white lab coat, a pocketful of pens, pop-bottle glasses, fuzzy hair, and buck teeth) where the system is presumably reduced to its smallest and simplest form (so that it can be more easily understood) and only one factor is allowed to vary at a time. Science spirals in toward what is presumably truth (or a close approximation) by testing (falsifying) all else. And if we do not use this method, then we are not card-carrying falsificationists.

According to this latter view, good science is doing, whereas history is just clever thinking; history is narrative, meaning that it consists of the "unique, unrepeatable, [and] unobservable" (Gould, 1986).[1] And if it can't be repeated, it isn't science. History, it is said, can demonstrate only tantalizing correlations (a relationship between two or more phenomena or sets of data), but it cannot demonstrate cause-and-effect. Such has been the "physics envy" of many biologists, geologists, and paleontologists, whose own disciplines are steeped in history that, ironically, history has been exorcised from discussion of the scientific method in introductory historical texts, or the scientific method is ducked altogether. Apparently because geology and other historical sciences are descriptive, not experimental, they are considered inferior. Perhaps in what is the ultimate act of capitulation to physics envy, most earth scientists in the mid- to late nineteenth century reluctantly accepted Lord Kelvin's calculation of the earth's age as being less than 100 million years, which was based on erroneous premises about the history of the Sun and the earth. Charles Darwin and other evolutionary biologists were at a loss to reconcile Kelvin's calculations, which were based on "hard" science and therefore seemingly correct, with their own conclusion—based on estimates of evolutionary rates and the geological record—that the earth must be substantially older. Indeed, the later discovery of radioactivity ultimately vindicated Darwin and his supporters (Eicher, 1976:13–19). A similar controversy has arisen more recently regarding the origin of cells based on "molecular clocks," which suggest a young age for early cell lines, versus evidence from the fossil record that indicates a much older origin for life (about 3.5 billion years ago) (Schopf, 1996; Smith and Littlewood, 1994; Doolittle et al., 1996).

Relationships between various physical and biological processes did not simply appear overnight. They have a history, and their origins

[1]Discussion based on Gould (1986), Frodeman (1995), and Salmon (1967).

and development are often contingent on what preceded them (Gould, 1986, 1989). Frodeman (1995:965) argues that in historical sciences, "the specific causal circumstances surrounding the individual entity (what led up to it, and what its consequences were) are the main concern. . . . In geology, the goal is not primarily to identify general laws, but rather to chronicle the particular events that occurred at a given location. . . . This means that hypotheses are not testable in the way they are in the experimental sciences. . . . Although the geologist may be able to duplicate the laboratory conditions of another's experiment . . . the relationship of these experiments to the particularities of Earth's history . . . remains uncertain." Geology involves the methodology of hermeneutics, which originated in the nineteenth century as a means of reconciling contradictions in the Bible. Later, the techniques were applied to other historical and legal documents to determine the original intent of the authors. In so doing, the interpreter must assign different values to different statements regarding their clarity or intent. In the same way, a geologist gives different weights to data (such as the stratigraphic continuity of an outcrop) based on the clues of past events and processes in a way analogous to how a physician makes a diagnosis or a detective builds a circumstantial case against a defendant (Frodeman, 1995:963). When a person dies, for example, we are not interested in the fact that the person died (which is what the reductionist law "all organisms die" tells us), but in the circumstances surrounding the death of the individual. Similarly, we are interested in the geologic history of a region or the lifespan of a particular species; we are interested in a central subject that serves as the basis for a historical reconstruction. In this sense, a geologist does produce a narrative, but not in a pejorative sense; rather, *narrative* means the explanation of a phenomenon by its integration into a larger story or context that gives meaning to the data (Frodeman, 1995). Indeed, Frodeman argues that this kind of reasoning is best suited for confronting the environmental problems that loom before us now and in the next century.

According to Gould (1986), Charles Darwin brought history into science because in his writings he established a methodology for historical sciences. First, is the uniformitarian argument: observable, gradual, small-scale changes may be extrapolated to much longer temporal scales to encompass the results in the geological record. Darwin asserted, for example, that the results of artificial breeding experiments "scale upward" to the genetic differences between local populations (demes), races, and species over geographic areas. In effect, he took an observation and asserted that it has reality beyond observation. He used a similar—but much more quantitative—approach to argue for the long-term effects of the lowly earthworm on the soil and landscape of England (Gould, 1986). This ontological approach, which emphasizes reality beyond human observation (as opposed to epistemology, or science that does not go beyond the consequences of direct observation; Allen and Starr, 1982) is an outgrowth of Charles Lyell's *Principles of Geology,* a highly influential book of Darwin's time (it was published in twelve editions between 1830 and 1875; Greene, 1982). Lyell's extreme philosophic view of geological processes as being slow and steady came to be regarded as the central dogma of the earth sciences: the principle of uniformitarianism, or "the present is the key to the past" (to which I and others would now tack on the addendum "sometimes"; see Bottjer et al., 1995; Martin, 1995a, 1996a, 1996b).

Not even James Hutton, whose observations laid the basis for this principle, shared such extreme views (Greene, 1982). Interestingly, however, reductionism seems to pop up in the earth sciences even as early as Hutton. "A given phenomenon . . . was not an isolated instance of . . . infinite possible combinations . . . but a decisive clue for the sharp detective of a *determinate outcome* predictable within narrow theoretical limits, a clue to the real history of the world, astonishing in its simplicity and beauty" (Greene, 1982:24). Ironically, Hutton, like Newton before him, believed that natural phenomena demonstrated the existence of a divine plan (Greene, 1982).

According to Greene, Hutton's approach was more deductive than inductive. In deduction, the conclusion of a logical argument contains no more than the initial premises, even though one or more of the premises may be absurd ("all squares are round; this is a square; therefore, this square is round"). An early typical example of deductive reasoning was to accept the Judeo-Christian god as the creator of the universe and its occupants, and to deduce what were thought to be the necessary consequences: The creation occurred only a few thousand years ago, all species have remained the same, and so on (Moore, 1993). Philosophers such as Sir Francis Bacon (1561–1626) found this approach repugnant and emphasized that one should begin with data, not faith (Moore, 1993). The conclusions of inductive arguments, on the other hand, contain more information than the premises (inductive is said to be ampliative) and the truth of the conclusion depends on further experience (observation, testing). Induction is an empirical approach; it uses observation and experiment to frame hypotheses.

Deductions (predictions) are then made from the hypotheses and tested (Moore, 1993). And if inductively derived generalizations—or deductions based on them—turn out to be incorrect, we seek new ones.

Unlike induction, deduction is nonampliative: It "purchases . . . truth preservation by sacrificing any extension of content" (Salmon, 1967:8). It is for this reason that scientists subscribe to the principle of the uniformity of nature. Such a principle is a statement that, when part of the premises of an inductive argument, makes an inductive argument deductive (nonampliative) in nature; synthetic statements are arrived at inductively, however, and their accuracy is indeterminate (Salmon, 1967).

Despite its name, then, the hypothetico-deductive method is not really deductive, and it was rejected by Popper, who emphasized that "good science" uses only deductive arguments to falsify hypotheses (Woodward and Goodstein, 1996). According to Popper, the more likely it is that a hypothesis can be falsified, the better it is. Hypotheses must run as great a risk as possible of being overturned because the more falsifiable a hypothesis is, the more it tells us; therefore, the more falsifiable a hypothesis is, the more it excludes extraneous possibilities and the greater the risk it runs of being false (Salmon, 1967). This approach differs from that assumed by most scientists (who attribute it to Popper): The more a hypothesis is corroborated by positive support, the more likely it is to be confirmed. Unfortunately, this often leaves too many hypotheses to explain the same phenomenon, and the hypothetico-deductive theorist will probably choose the most probable one, whereas Popper would pick the least likely one because of the inverse relationship between falsifiability and probability. According to Popper, a highly falsifiable hypothesis that is stringently and repeatedly tested and left unfalsified becomes highly corroborated, which is *not* the same thing as being confirmed. Popper considered his method to be strictly deductive; nevertheless, inductive argument still creeps in because without it, conclusions (hypotheses) would only confirm the premises (observations), and science would "amount to [no] more than a mere collection of . . . observations and various reformulations thereof" (Salmon, 1967:24).

Salmon (1967:131) goes one step further and suggests that "we . . . assess the probability that our observational results would obtain even if the hypothesis under consideration were false. . . . For strongest confirmation, this probability should be small." In statistical tests, the probability of accepting a false hypothesis (or more properly said, not rejecting it), is referred to as β error. The smaller the β error, the greater the power (sensitivity) of a statistical test, and the more powerful the test, the more likely it is to reject a hypothesis. The converse—rejecting a hypothesis when it should have been accepted—is called α error.

But what if we have only the *results* of experiments available to us, as is the case with the stratigraphic (fossil) record? Or what if the processes or entities involved are not observable within our frame of reference (Woodward and Goodstein, 1996)? We are not, as the "hard" science sycophants would have us believe, relegated to the "soft" sciences and constrained to merely observe and describe (both of which may be subtly influenced by our previous experiences and perceptions; see Woodward and Goodstein, 1996) and to extrapolate from human spatio-temporal scales. Based on our observations, we can recognize stages of processes, arrange them into hypotheses, and then test them. If this sounds unscientific, think about how the stages in the evolution of stars, from birth to death, are treated as a given in introductory astronomy texts. Darwin used this approach to arrive at an explanation of coral atolls. He recognized three types of reefs (fringing, barrier, and atoll) and linked them together, by a common historical hypothesis, as sequential stages of a single process: subsidence of islands. Although rival hypotheses emerged, Darwin's was ultimately tested and verified by later drilling of coral platforms (Gould, 1986).

It is important to emphasize that Darwin's hypothesis depended on his taxonomy (classification) of the entities (reefs) he wished to explain. Taxonomy is far from being a has-been of historical science, suitable only for deluging us with unimportant publications to be counted in one's vita (although, unfortunately, this is often the case). Rather, "in a profession more observational and comparative than experimental, the ordering of diverse objects into sensible categories becomes a sine qua non of causal interpretation" because it represents a causal ordering (Gould, 1986).

We may also infer history strongly from single objects by looking at anomalous features or imperfections. The odd layers of sediment and iridium found by Walter Alvarez et al. (1980) near Gubbio, Italy, ushered in a whole new view of the kinds and rates of processes that affect this planet (see chapter 8) and helped to break Lyell's view that had virtually straitjacketed geology for well over a century. Nor is a single entity beyond the hypothetico-deductive method using the historical record. Although each singularity (note the bow again to physics) is indeed unique, collections of singularities may exhibit certain nomothetic (general or universal) relationships that can be predicted and tested. If iridium is indeed associated with extraterrestrial impacts and extinction, then perhaps it is associated with other extinctions as well; if so, then more than the end-Cretaceous extinction may be attributable to impact. If that is the case, then impacts are much more frequent than previously imagined and therefore play a much greater role in determining the course of this planet's evolution (I return to this topic again in chapter 8). Moreover, there may be iterative patterns that couple data from diverse areas and constrain our simultaneous explanation of them, so that only one explanation is plausible.

There are patterns and processes, then, that are observable only in the geological record, and the scales we use to measure phenomena are critically important in determining the processes we observe (the subject of chapter 1). It was not until fairly recently, for example, that seafloor spreading rates could be accurately calculated with sophisticated techniques; before that time, spreading rates were so slow (a few centimeters per year, on average) as to be unmeasurable, so naturally the continents did not drift about, but remained fixed throughout their history. Anyone who lived through the often acrimonious debates between "fixists" and "drifters" remembers the intensity and the animosity between proponents of both sides. As another example, because species evolve far more slowly than is detectable to human scales of observation, many early investigators concluded that all species *must* have been created at once by a special creator.

This is not to say that historical approaches are perfect. The emphasis on observation and description in historical sciences has caused us to focus on causal processes that are perceptible on human timescales, and therefore presumably explainable in terms of present-day processes, just as Lyell propounded. Darwin's upward-scaling of short-term processes to longer timescales, for example, is by no means unequivocal and it should be an important consideration not only in geology, but also in other fields such as ecology, which should also be concerned with history and contingency. But to me, imagining—and documenting—what those patterns and processes are is the purest pleasure of being a paleontologist and working on this planet's history.

1

Scale, Measurement, and Process: An Introduction

Tangibility narrows . . . vision through prejudice. —*Allen and Starr*

SCALE AND MEASUREMENT

Everyone is familiar with the term *scale,* whether it be spatial (meter sticks, for example) or temporal (minutes, hours, and days, up to the billions of years of the geologic timescale). What is not commonly appreciated is that the choice of scale dictates the accuracy of the resulting measurement. In most circumstances this is unimportant, and the scale used is chosen for its convenience. Whether a distance (length) is measured in centimeters or kilometers, say, depends on the size of the object relative to the convenience of the scale of measurement. If we were to measure the coastline of Great Britain (or any other land mass or other natural object, such as a river), for example, we would undoubtedly choose the kilometer as the scale of measurement instead of centimeters, or we would spend many lifetimes measuring the distance. Similarly, if we were to measure a sine wave with a wavelength (ℓ) of several kilometers, we would not measure ℓ with a single meter stick; if we used a meter stick, we would only detect ℓs of about the same length as the meter stick, and we would probably not detect the sine wave because our scale would be too short (figure 1.1).

What if we were to measure the duration of cyclic climate change

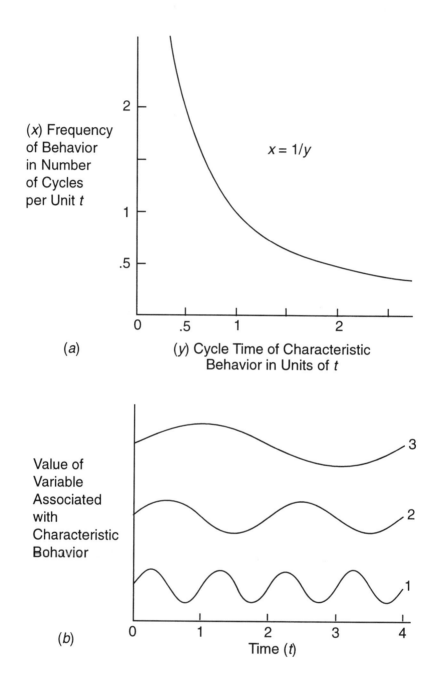

(x) Frequency of Behavior in Number of Cycles per Unit t

(y) Cycle Time of Characteristic Behavior in Units of t

$x = 1/y$

(a)

Value of Variable Associated with Characteristic Behavior

(b)

Time (t)

through time? If each cycle were of several million years' duration, we would not even see any change if our scale was one human generation. The environment would appear to be fairly constant to us because the duration of observation is much too short to detect long-term climate change. The same would hold true for a long-term change in a particular direction: secular change. What we detect or observe, then, depends on our scale of observation.

Let's look at the coastline example more closely.[1] If we were to measure the coastline of Great Britain using a measuring stick of many hundreds of kilometers length, we would get a shape like the left-hand one in figure 1.2, whereas we would get a more familiar-looking coastline using a 1-km measuring stick (figure 1.2, *right*). In other words, choice of scale determines the resultant shape. As the scale length (ℓ) decreases, the number of times we must use the scale to measure the object increases. In fact, if N = the number of times the scale must be used to measure a particular object, then

$$F = \lim_{\ell \to 0} \left[\frac{\log N(\ell)}{\log(1/\ell)} \right] \qquad (1.1)$$

[1]Much of the following discussion is based on a brief, nonmathematical treatment in Allen and Hoekstra (1992:54–66) and the simple but more mathematical treatment of Sugihara and May (1990). Chapter 1 of Kellert (1993) also provides a brief but lucid introduction to fractals and chaos theory.

FIGURE 1.1 Inverse relationship between cycle time and frequency (a) for three different cycles (b). If the scale of measurement is sufficient to measure the length of cycle 1, it is too short to measure the true frequency of cycle 3. Conversely, if the scale is long, cycle 3 is detected, but not cycle 1. In the latter case, certain points of cycle 1 may be incorporated into cycle 3 and produce error known as aliasing, in which higher frequencies are incorporated into lower frequencies (Davis, 1986). *From T. F. H. Allen and T. B. Starr,* Hierarchy: Perspectives for Ecological Complexity. © *1982 by the University of Chicago. Reprinted by permission.*

FIGURE 1.2 The detail and length of the coastline of Great Britain increases as the scale of measurement decreases. For infinitesimal line segments, the coastline is infinite in length. *From* Toward a Unified Ecology *by Timothy F. H. Allen and Thomas W. Hoekstra. Copyright © 1992 by Columbia University Press. Reprinted with permission of the publisher.*

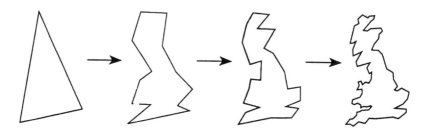

(Kellert, 1993:17). In other words, if we use a scale of smaller and smaller length, the ratio of log $N(\ell)$ over log $(1/\ell)$ will converge toward a limit called F (the fractal dimension) that describes the shape of the object. Fractal dimension refers to *fractional* dimension, a term coined by its originator, Benoit Mandelbrot. Unlike the simple three-dimensional mind set that is taught in elementary geometry, fractal objects possess dimensions that are not whole numbers. A line has a dimension of 1 and a plane a dimension of 2. In the case of the plane, the line "fills" the entire plane (Brownian motion or random walk), but a fractal object with $F = 1.26$ (such as Koch's snowflake; figure 1.3), for example, lies somewhere between a line and a plane.

F, then, is a measure of the complexity of the object's shape. Note that in equation 1.1, F is represented by the ratio of two logarithms. A log–lot plot of ℓ versus the total measured perimeter (P) length of the object yields a straight line for a fractal object (figure 1.4). The slope (m) of the line is negative (slopes downward to the right) because the longer ℓ is, the shorter the total estimated perimeter (or, conversely, the shorter ℓ is, the greater the total perimeter measured). Complex shapes have a steeper slope than simpler ones because the measured perimeter of a complex object increases more as ℓ decreases than does the perimeter of a simpler object.

To determine F for an object, one can measure P as ℓ is decreased, plot log P versus log ℓ, and then plot (regress) a line through the points (see Sugihara and May, 1990, for other methods of calculation). The equation for this straight line is

FIGURE 1.3 Koch's snowflake as an example of self-similarity. In successive diagrams, the midpoint of each line segment is pulled out to produce new line segments equal in length to the original. Successively smaller portions of the snowflake (move from right to left in the diagram) are small-scale versions of the entire snowflake. *From* Toward a Unified Ecology *by Timothy F. H. Allen and Thomas W. Hoekstra. Copyright © 1992 by Columbia University Press. Reprinted with permission of the publisher.*

Self similar pattern

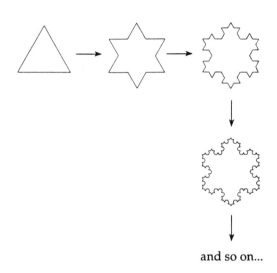

and so on...

FIGURE 1.4 Log–log plot of the length of the measurement scale (ℓ) versus total perimeter (P) of an object. Straight line segment indicates that the object is fractal, and the slope of the line (m) indicates whether it is simple or complex. *From* Toward a Unified Ecology *by Timothy F. H. Allen and Thomas W. Hoekstra. Copyright © 1992 by Columbia University Press. Reprinted with permission of the publisher.*

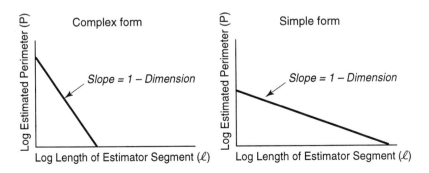

$$\log P = (1 - F)(\log \ell) + \log K, \qquad (1.2)$$

where the slope $(m) = 1 - F$ and $\log K$ is a constant (y-axis intercept, $\ell = 0$). This is just the familiar equation for a line $(y = mx + b)$ in a different guise. Rewriting equation 1.2 yields

$$P = K\ell^{1-F} \qquad (1.3)$$

(Sugihara and May, 1990); in other words, $\log (K\ell^{1-F}) = \log (\ell^{1-F}) + \log K = (1 - F) \log \ell + \log K$. The total perimeter (P) of a fractal object is equal to the number of sides (N) times the length of the side (ℓ). For a fractal object, as ℓ approaches 0 (and F approaches the actual fractal dimension), the perimeter length approaches an asymptote, K (y-axis intercept; figure 1.4). Each side has a fractal-dimensional length (ℓ^F), so the number of sides N equals the total true perimeter (K) divided by the fractal-dimensional length of each side (ℓ^F):

$$N = \frac{K}{\ell^F} \qquad (1.4)$$

(or $N = K\ell^{-F}$). Because $P = N \times \ell$, P may be expressed in more general form as

$$P = K\ell^{-F}(\ell), \qquad (1.5)$$

where $\ell^{-F}(\ell^1) = \ell^{1-F}$. Now, if we take the logarithm of both sides, we return to equation 1.2, which is why the slope of the line is $1 - F$.

One implication of the straight-line plot of figure 1.4 is that the fractal dimension (F) is constant; the complexity of the object is scale-invariant, and all fractal objects are said to be self-similar. In the case of Koch's snowflake (figure 1.3), for example, successively larger areas of the curve are identical to smaller portions and vice versa. This does not mean that natural objects display such strong regularity. Natural objects are much more likely to display a *statistical* self-similarity; the similarity at different scales is quite strong, but not exactly invariant like artificial patterns such as the Koch snowflake (Peak and Frame, 1994). In other words, for natural fractals the straight line of figure 1.4 is quite likely to be plotted through a cloud of points rather than a series of points lying exactly on a straight line. Natural (statistical) fractals occur all around us: coastlines (from grains of sand to coasts viewed from space), ferns, trees, rivers, landscapes, surfaces of boulders, and perhaps portions of stable isotope or fossil abundance curves, among other examples. Fractals are not found everywhere, however; in fact, given the resolution of most measuring instruments, F probably equals 1 for most natural objects (Sugihara and May, 1990).

HIERARCHY AND SCALE

Do all fractal objects have a constant fractal dimension? No. In fact, the fractal dimension typically varies, and this is where process comes into play. The scale of measurement often dictates the process measured. The dimensions of natural fractals typically vary, so the processes that produce the object (the thing measured or observed) also vary according to scale (I use the word *object* loosely; see Salthe, 1985, for precise definitions of the words *object, entity,* and *thing*). If the fractal dimension does change with change in scale, then the implication is that there is a hierarchy of processes involved that produced the object (Sugihara and May, 1990).

We are all familiar with hierarchies. The scheme of biological classification (species, genus, family, order, class, phylum, kingdom, etc.) is an example of a nested hierarchy of entities (smaller boxes are contained within larger boxes); as discussed in chapter 5, a dispute rages between evolutionary biologists about the origins of taxa within these categories. The rock units member, formation, and group; the ecological hierarchy cell, tissue, organ, organism, population, and community; and the genealogical hierarchy codon, gene, organism, deme (local population), species, and monophyletic taxa are also nested hierarchies (table 1.1; Eldredge and Salthe, 1984). Salthe (1985) gives an extensive justification of the ecological and genealogical hierarchies. Notice that the level *organism* appears in both ecological and genealogical hierarchies (we return to this important point in chapter 7). Also, hierarchies are not necessarily nested.

Although a hierarchy may consist of objects, their hierarchical arrangement implies process. Two shapes of different complexity are shown in figure 1.5, both of which exhibit a sharp change in slope of the log–log plot. The object on the left exhibits a great deal of complexity (high fractal dimension) at small scales of measurement (strong negative slope of the line), whereas the object on the right is quite simple (low fractal dimension, less steep slope). At larger measurement scales, however, the pattern of the shape on the left is reversed. If our measurements pertained to land use patterns, then the left-hand example would probably represent anthropogenic activity, which tends to straighten things (such as roads and the boundaries of fields and forests) for purposes of record-keeping, whereas the right-hand example probably represents the effects of natural topography (valleys and hillsides, trellis drainage patterns, and so on).

Do processes that operate over small areas necessarily scale upward to much larger areas? In other words, are processes that act over short and long distances self-similar? The answer is that it depends. In the case of topography, the answer is no. In the case of sedimentary rocks, bedding or stratification varies according to grain composition and size, which vary according to the source area and energy regime. Boundaries between larger rock units (formations) may also indicate changes in source area and energy or they may suggest changes in the site of deposition, or both. Jackson (1991), for example, distinguished between two basic scales in coral reef ecology: quadrat and landscape. In the quadrat view (a quadrat is a square, usually measuring about 1 m × 1 m, within which ecologists measure such variables as percentage species cover and diversity), things such as relative species abundance and diversity fluctuate unpredictably. This observation led Connell (1978) to explain diversity in terms of his intermediate disturbance hypothesis, which is discussed further in chapter 8, and appears to have been preceded by a similar idea proposed by Horn (1975; see also May, 1994:4). On the other hand, at the larger landscape scale, the relative proportion of habitat in each type of disturbed state remains fairly constant or predictable. Jackson (1991) concluded that at the small (quadrat) spatial scales, short-term biotic interactions and

TABLE 1.1 Genealogical and Ecological Hierarchies

Genealogical Hierarchy	Ecological Hierarchy
Codons	Enzymes
Genes	Cells
Organisms	Organisms
Demes	Populations
Species	Local ecosystems
Monophyletic taxa	Biotic regions
(Special case: all life)	Entire biosphere

Source: Eldredge and Salthe, 1984.

Note: Hierarchies are arranged so that the only levels (holons) shared by the two hierarchies—organisms—are aligned. Resulting alignment of other holons does not imply identity or equivalence.

16

larval recruitment are important determinants of coral distributions, whereas patterns of resource availability, routine physical processes (such as wave and current action), and major disturbances select for traits that result in increased predictability of species distributions on larger spatial scales (see table 1.2).

With respect to temporal scales, the answer appears to depend on the process and the duration of time involved. In the case of land use activities, the answer again appears to be no. Human activities operate over much shorter periods of time than geological (geomorphological) ones. Perhaps this is why environmentalists have had such a hard time convincing politicians that humans do in fact have a significant and often detrimental impact on the earth and its biota; humans don't live long enough to experience much of the long-term impact of many generations of our species. Similarly, Jackson (1991) concluded that the

FIGURE 1.5 If the slope of the line in a log–log plot of the measurement scale versus total perimeter changes, then the fractal dimension (and complexity) changes. These particular figures are *not* fractal, but are used instead to indicate that a large- or small-scale pattern can exhibit different degrees of complexity. *From* Toward a Unified Ecology *by Timothy F. H. Allen and Thomas W. Hoekstra. Copyright © 1992 by Columbia University Press. Reprinted with permission of the publisher.*

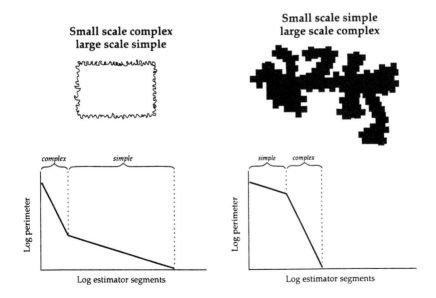

temporal scales of processes acting on coral reefs varied according to spatial scale (table 1.2). Aronson (1994), on the other hand, concluded that many ecological and evolutionary processes (such as the effects of predation on morphology, abundance, and distribution of prey through space and time) scale upward temporally, whereas Eldredge (1995) argued forcefully that the fossil record contains patterns indicative of evolutionary processes and rates that cannot be scaled upward from, say, population biology alone. Do processes of mass extinction scale upward from local (ecological) causes or must they always be caused by some unusual and catastrophic agent? Does it matter, for example, if the dinosaurs died out with a bang or a whimper (Gould, 1992)? If we were to go back in time during a mass extinction, would we be able to tell that one was occurring based on the limited temporal perspective of a human lifespan? Probably not. But then how gradual is gradual? How fast is catastrophic?

Suffice it to say that upward (spatial and temporal) scaling of geological and biological processes must be examined on a case-by-case basis. Before we go on, though, perhaps we should take a closer look at the concept of hierarchy.

TABLE 1.2 Ecological Processes and Disturbances on Caribbean Coral Reefs at Different Spatial and Temporal Scales

Process	Spatial Extent	Duration	Frequency
Predation	1–10 cm	Minutes–days	Weeks–months
Damselfish gardening	1 m	Days–weeks	Months–years
Coral collapse (bioerosion)	1 m	Days–weeks	Months–years
Bleaching or disease of individual corals	1 m	Days–weeks	Months–years
Storms	1–10^2 km	Days	Weeks–years
Hurricanes	10–10^3 km	Days	Months–decades
Mass bleaching	10–10^3 km	Weeks–months	Years–decades
Epidemic disease	10–10^3	Years	Decades–centuries
Sea level or temperature change	Global	10^4–10^5 Years	10^4–10^5 Years

Source: Jackson, 1991.

THE CONCEPT OF HIERARCHY[2]

Hierarchies consist of discrete levels called holons. Each holon has three aspects: (1) its interior, which consists of (2) its parts (which may in turn be separate holons with their own parts), and (3) its surrounding environment (which may be another holon surrounded by its environment). Thus holons are both parts and wholes simultaneously (Salthe, 1985). A taxonomic holon, for example, is a level that contains other objects (such as the species of a genus) and is in turn subsumed by a higher taxon (a family of genera in the taxonomic hierarchy). Imagine watching a football game from a dirigible above (see also Allen and Starr, 1982:figure 8.9). If one is too far away, the game may be obscured by significant surfaces (clouds or the walls of the stadium) that represent the surfaces of holons. Now imagine being slowly lowered by a rope into the stadium holon, viewing the crowd and game as a single entity. As we penetrate the cloud holon, it is the stadium holon (crowd and game) that fills our field of vision. And as we near the ground, we penetrate the game holon; as we land in the midst of the game, we see only a few players at any one time. If we continued further, we might become one of the players ourselves, viewing the game from his eyes alone. Or perhaps we would be unfortunate enough to pass through the wall of the football holon, instead, in which case our view might be completely black because we're inside the football (or because a 300-pound player has fallen on us, the football). Note that in choosing between the inside of a football player or the football, we have constructed a hierarchy with branches (non-nested); the upper holons of one branch do not constrain the lower holons of another branch, nor are the higher holons derivable from the lower ones, as they are in nested hierarchies.

A holon, then, is an integration of parts, but this definition says nothing about how many parts there are or the criteria to determine what is and what is not a holon. The easiest (and most arbitrary) way to define a holon is to let it be defined by the observer. Boundaries (discontinuities) perceptible to the human sensory apparatus are likely to appear fairly distinct because there are likely to be fairly abrupt changes in characteristics or phenomena across the boundaries; other-

---

[2]This discussion is distilled from Allen and Starr (1982) and Salthe (1985). I recommend both works, although they are not easy reads. See also Ahl and Allen (1996) for a briefer and easier discussion of hierarchy.

wise, they would not be recognized as boundaries. Perhaps this is why the holon *organism* is so prominent in biological and paleontological thought (Eldredge and Salthe, 1984; Salthe, 1985). As we move further up the ecological hierarchy, though, ecosystems and biotic provinces become increasingly unfamiliar to us, and, unfortunately, remain vague and abstract in part because they are so far removed from our own "organismic" experience, and perhaps also because higher holons have had less time to evolve and become more discrete (see chapter 7; see also Salthe, 1985). Boundaries, then, may partly be a function of our ignorance (Salthe, 1985).

The process of reifying one level—or asserting that an observation, which is based on human perception, has ontological reality as a holon (the word bears the same root as *paleontology* or the study of ancient beings)—is fraught with danger because the boundaries or objects are self-referential; they are based on our own experience. There may be other holons, just as in the football game, that are invisible to us from our standpoint. An ecologist who describes disturbance to an ecosystem (say a forest fire) over a time span of a few years may be correct that disturbance has occurred, but in the longer term—beyond the ecologist's period of observation—fire may be a normal, even necessary component (for seed germination) of the ecosystem. The disturbance has actually become incorporated into the ecosystem over long intervals of time and is not really a disturbance at all over sufficiently long timescales.

Salthe (1985) lists several criteria for distinguishing holons. Besides boundaries, he suggests the criteria of scale (something must be bigger or smaller than something else) and integration (which we have already mentioned in terms of holons). If the holon is repeatedly recognized using different techniques, then it is robust (Salthe, 1985, delves into these topics from a philosophical approach; Allen and Starr, 1982, suggest multivariate statistical methods that can be used to detect holons). A holon also exhibits spatiotemporal continuity: Whatever we recognize, it is sufficiently stable to persist over some area and last for some interval of time. Holons also have a history. This is of fundamental importance to historical scientists. The holons have unique properties or were determined by unique configurations of historical contingency. Hierarchy works by recognizing differences (history); it is what Salthe (1985) calls an idiographic approach, which emphasizes particularities. On the other hand, reductionist science works by using observational regularities or similarities ("laws") discovered by comparing measurements (Salthe, 1985), and is a nomothetic approach because it seeks general or universal laws. (It is partly for this reason that different scientific disciplines and subdisciplines within a field can be pursued independently.) The rates of the processes studied using idiographic and nomothetic approaches may not interact directly (Salthe, 1985), thereby isolating the disciplines from one another.

Hierarchies may also be viewed as systems of constraint. When examining a holon we must consider the holon immediately above and that immediately below (which forms the constituent parts of the holon in question). Higher (larger) holons tend to constrain the behavior of their constituent lower holons because the higher holons provide the environment (boundary conditions) within which lower holons must operate; conversely, lower holons provide the initiating conditions or possibilities that, depending on the boundary conditions, may or may not be realized through the process of adaptation as "aptations" (the state of relatedness of a system to its environment; Eldredge and Salthe, 1984; Salthe, 1985). The greater the number of boundary conditions, the fewer the possibilities that are realized. The communication between the holon being observed (the focal level) and its components and environment is strongest. Holons located farther away than this triad typically exert much less constraint on the focal level. If this were not so, then the entire hierarchy would collapse (or never have developed at all) because of mass transitivities across boundaries between holons (Salthe, 1985). The heat capacity and moisture from polar seas serve as initiating conditions (that is, they make it possible to have glaciers), but without the proper tilt of the earth (boundary conditions), which affects solar insolation, there would be no glaciers (Salthe, 1985). *Higher-level constraints produce boundaries that are historical in nature, whereas lower-level processes act in a more lawful manner* (Salthe, 1985).

Thus constraint consists of the product of two factors: the amount of information exchanged between holons and the asymmetry of information exchange. Whereas the asymmetry between lower and higher holons is quite large, the behavioral frequency of a lower holon is short enough that it has little or no communication with a much higher holon, and a much higher holon exerts little or no effect on the lower holon, so the product is 0. At the other extreme are holons that operate at the same level but in other compartments (sister holons). Here there is a large amount of information exchange (through another higher level), but no asymmetry, so the product is again 0, and they exhibit no constraint over one another.

Another way to define holons is to view a hierarchy as a continuum of natural frequencies (figure 1.6). As more is included within the boundaries of the system being studied, more and more components, which may be thought of as behaving at different frequencies, are also included, so that the spectrum of frequencies of the entire system (hierarchy) becomes continuous. Nevertheless, if the frequencies of different processes are sufficiently different (but not *too* different) they interact to produce a "standing wave" or boundary recognizable to an observer (Salthe, 1985). Higher holons behave more slowly (have lower frequencies or cycle times of undisturbed behavior) than lower holons, which have higher frequencies. If disturbed, the "return or relaxation times" (resilience) of higher holons is also slower than those of lower holons. Chemical weathering of continental rocks, for example, results in a long-term (low-frequency) control on atmospheric CO_2 that occurs over tens to hundreds of millions of years; on the other hand, changes in rates of marine and terrestrial photosynthesis during the Pleistocene ice ages exerted control over atmospheric CO_2 on much shorter timescales (higher frequencies) of *only* thousands of years because ocean circulation rates were on the order of a thousand years

FIGURE 1.6 Frequency and constraint of holons. Constraint is minimal between holons of nearly equal or very different frequencies. *From T. F. H. Allen and T. B. Starr*, Hierarchy: Perspectives for Ecological Complexity. *© 1982 by the University of Chicago. Reprinted by permission.*

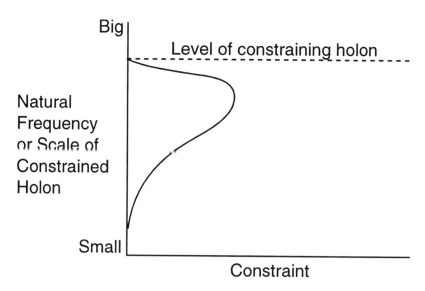

(chapter 6). These two holons may be recognized because the ratio of their two frequencies is so large. The higher holon behaves so slowly, however, that it appears constant to the lower (faster) holon, which appears quite noisy to the higher holon (as discussed in chapter 7, there may also have been a secular increase, or increase through geologic time, in weathering rates). On the other hand, changes in ice cover of the continents, and associated changes in ocean circulation rates, are rapid enough that they constrained photosynthetic rates and atmospheric CO_2 over much shorter durations (thousands of years) during the Pleistocene than did continental weathering. Note that although these holons are not necessarily nested, they are still in some form of communication with one another through their effect on CO_2 levels.

In this way, constraining holons ("governors") integrate (filter) the signal (information) from lower holons so that the signal is averaged or lagged at the level of the higher holon (the signal is actually a statistical average that appears constant at the higher level). Constrained holons also filter signal from higher holons, but over such a short period that the message from the governor is read immediately and it too appears constant to the constrained holon. Thus the position of a holon in a hierarchy, which is another way of viewing its scale, is determined by its ability to process information. But the lagging (filtering) of information is not readily apparent from our human frame of reference. A motion picture, for example, consists of individual frames lagged (filtered) twenty-four times a second to produce what appears to the naked eye to be a continuous sequence of images.

This brief introduction to hierarchy has been largely abstract. It is not my intent to review the subject exhaustively (others have done so before me; see Allen and Starr, 1982, and Salthe, 1985), but to introduce the concept of hierarchy, as I return to it repeatedly throughout the book. It is better at this point to shift focus to the nature of the stratigraphic record. This historical record consists of a conglomeration of cycles (occurring at different frequencies), secular trends, singularities (events), and hierarchies, all superimposed on one another. The record is by no means complete, but it is our job to sort it out because it appears to contain evidence of processes not perceivable on human timescales.

2

The Nature of the Stratigraphic Record: Curds and Whey

Nothing but time. —*James Hutton,* Theory of the Earth

Time is Nature's way of keeping everything from happening at once.
—*Chicago restroom graffiti*

In this chapter, I discuss the relationships between time, rock, and time–rock units. Although students are trained in the meaning these terms, I suspect much of it remains a mystery, and that the relationship of these units to the nature of the stratigraphic record is often misunderstood. As will be evident in later chapters, stratigraphic (temporal) resolution is critical to establishing any cause-and-effect preserved in the rock record.

TIME, ROCK, AND FACIES

How do we know whether an outcrop or a fossil is, say, Ordovician, Mississippian, or Cretaceous in age?[1] Just a hundred or so years ago, these names did not even exist. Now geology students memorize epithets to remember them and the rest of the geologic timescale. But what do they mean? To answer this question and move forward, we must back up and see how these terms came into being.

The earliest timescales were based on superposition (younger rocks lie on top of older rocks) and lithology (rock type). Each basic rock type represented a distinct phase in the earth's history. (Fossils were

[1]Eicher (1976) and Berry (1987) provide excellent summaries of the development of the geologic timescale, on which much of this discussion is based.

24

seen by some as the work of the devil or were ignored altogether.) Thus the earliest rocks were called Primary and consisted of crystalline rocks with metal ores (Ganggebirge, or ore mountains). Secondary rocks came next and were stratified (layered) and contained fossils (Flötzgebirge, or layered mountains). On top of Secondary rocks came Tertiary rocks, which also contained fossils but were often poorly consolidated (alluvium, also called Angeschwemmtgebirge, Aufgeschwemmtgebirge, or Neues Flötzgebirge). Those already familiar with the timescale will recognize that the term *Tertiary* is still with us (figure 2.1), although it doesn't mean the same thing that its originator (Giovanni Arduino) intended.

Similar timescales were being developed by other workers in Europe and Russia about the same time, which led them to conclude that these units were geographically widespread. Rocks tend to repeat themselves laterally (spatially) and through time (Walther's law), however. Sandstones, granites, and other rocks have formed through much of the earth's history, not just during a particular phase, so that the Primary, Secondary, and Tertiary could not really be reliably matched

FIGURE 2.1 Development of the geologic timescale. Authors and date of initial description of units are indicated. *Redrawn from Mintz, 1981.*

Early Subdivision				Modern Usage			
Arduino 1760	Lehmann 1756 Füchsel 1760-1773	Werner ca. 1800	English Equivalents	Eras	Periods	Epochs	Alternate Periods
Volcanic Alluvium	Angeschwemmt-gebirge	Aufgeschwemmt-gebirge or Neues Flötzgebirge	Alluvium	Cenozoic Phillips 1841	Neogene Hoernes 1853	Holocene Pleistocene	Quaternary Desnoyers 1829
Tertiary			Tertiary			Pliocene Miocene	
					Paleozoic Naumann 1866	Oligocene Eocene Paleocene	Tertiary Arduino 1760
Secondary	Flötzgebirge	Flötzgebirge	Secondary	Mesozoic Phillips 1841	Cretaceous d'Halloy 1822		
					Jurassic von Humboldt 1799		
					Triassic von Alberti 1834		
					Permian Murchison 1841		
					Pennsylvanian Williams 1891	Carboniferous Conybeare and Philips 1822	
				Paleozoic Sedgwick 1838	Mississippian Winchell 1870		
Primitive	Ganggebirge	Übergangsgebirge	Transition		Devonian Murchison, Sedgwick 1839		
					Silurian Murchison 1835		
					Ordovician Lapworth 1879		
					Cambrian Sedgwick 1835		
		Urgebirge	Primary		Precambrian		

(or correlated) according to time much beyond the areas in which they were originally described.

Rocks can be equated with time only through use of the two "f" words of geology: *formation* and *facies*. A formation is a mappable unit of rock based on lithology; it is a descriptive term only, and no interpretation is involved in its recognition. The term *facies*, as originally used by Swiss geologist Armand Gressley in 1838, refers to the environment of deposition; different environments exist at the same time and the same types of environments exist at different times (Walther's law). So what is the connection between formations (rocks), facies (environments), and time?

Let's take the standard example of the transgressive (rising sea level) sequence of the Cambrian Tapeats Sandstone (oldest), Bright Angel Shale, and Muav Limestone (youngest) found at the base of the Grand Canyon (figure 2.2; the names come from the predominant rock type or lithology and the place, or type locality, where they were first described). For reasons of energy distribution (waves and currents) and sediment supply, sandstones are found closest to shore, shales farther away (finer particles settle to the bottom only in quieter water), and limestones farthest offshore (a lack of sediment input results in rocks that consist mainly of shells). All three facies (or lithofacies, because they're based on lithology) are present at the same time. Suppose the sea level rises as time passes. The sandstone–shale–limestone sequence will track sea level and move landward. If we were to equate rock type directly with time at site A or B, then, based on superposition, we would have "sandstone time" followed by "shale time" and then "limestone time." Based on this interpretation, *all* the sandstone was deposited during the same interval of time, then all the shale, and then nothing but limestone. In other words, rock equals time. Period. But that is not what happened in our example of transgressive facies. All three lithofacies coexisted, so the time lines must be drawn *across* the formation boundaries, not *parallel* to them. To equate rock with time at the outcrop scale is no more satisfactory than at the much higher levels of Primary, Secondary, and Tertiary.

Early originators of the timescale equated rock with time because the significance of fossils was unknown to them. Some, such as Abraham Werner, held quite extreme views, believing that *all* rocks of the same lithology constituted a formation. This was because, according to Werner, all rocks of the Primary layer (which included igneous rocks such as granites; figure 2.1) were deposited from a global ocean (Berry, 1987).

FIGURE 2.2 Facies relationships illustrating the difference between rock, time, and time-rock units. If rock (lithology) is equated with time, then all the sandstone was deposited first, then all the shale, and then all the limestone. However, timelines indicate that these three lithofacies (and the environments in which they were deposited) *coexisted*, and tracked the sea-level rise illustrated in the diagram. *From* Historical Geology *by L. W. Mintz, © 1981. Reprinted by permission of Prentice-Hall, Inc., Upper Saddle River, NJ.*

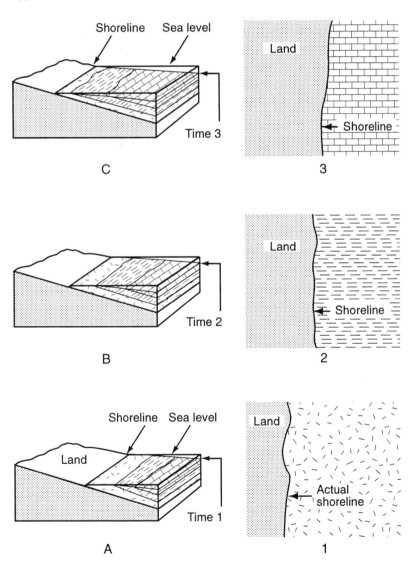

To be sure, if one were to *correlate* (determine the age equivalence or contemporaneity of the rocks) based on rock type alone in our example, there would probably not be any serious errors *as long as the correlation was done over short distances* (a procedure that geologists call lithocorrelation or "walking the outcrop" because it is done within a local area). At the other (hypothetical) extreme, if one correlated from the Grand Canyon clear across the continent based on lithology alone, very serious errors would result! In essence, this is what the originators of the earliest geologic timescales tried to do. Younger rocks lay on top of older rocks (the principles of superposition and original horizontality developed by Nicolaus Steno in the late 1600s) and lithology in turn was indicative of time.

William Smith enters the picture in 1815. Smith was an English surveyor, and because of his vocation, he had ample opportunity to observe rocks and fossils over large areas. He noticed that fossil assemblages always occurred in the same stratigraphic (superpositional) sequence, which led to his recognition of the principle of faunal succession: Life, unlike rocks, has changed through time and tends not to repeat itself, which is a result of biological evolution (sometimes called Dollo's law). Bear in mind, however, that Charles Darwin's opus *On the Origin of Species* was not published until nearly a half century later, in 1859.

Smith's observations were responsible for the first geologic map of England, which was based on his subdivision of rocks belonging to the Secondary and Tertiary portions of earlier timescales. The Primary period is essentially what we now call Precambrian, and although it is fascinating because it records the origin and development of the earth and its early life in its two main subdivisions, Archeozoic (ancient life) and Proterozoic (first life), most paleontologists concentrate on the Secondary and Tertiary—about the last 600 million years—because it is the time of apparent life (Phanerozoic), when fossils are typically abundant. By the middle of the nineteenth century, most of the geologic periods, which were each characterized by a particular kind of fossil (animal or plant) assemblage (figure 2.1) had been recognized in Europe, although the Carboniferous (coal-bearing) units were later subdivided into the Mississippian and Pennsylvanian in North America (the basic units can be remembered by the mnemonic "Cold Oysters Seldom Develop Many Precious Pearls; Their Juices Congeal Too Quickly").

Thus the geologic periods are based on major changes in the earth's biota through time. A period is a particular kind of time unit, whereas the corresponding term *system* is a type of time–stratigraphic unit (the rocks deposited during a period). Periods (systems) can be further subdivided into epochs (series), stages (ages), and chrons (zones). A stratigraphic facies is also a kind of time–stratigraphic unit because it is bounded above and below by time. As we saw previously, the distinction between rock and time units is one of the most fundamental in all of stratigraphy (and in all of geology), and the distinction between time and time–stratigraphic units is no less significant.

THE HIERARCHY OF THE TIMESCALE

The subdivisions of the timescale can tell us a great deal about process. The smallest practical subdivisions of the scale are biostratigraphic zones, usually encompassing 1 to 2 million years. German paleontologist Albert Oppel developed the process of biostratigraphic zonation in Europe in the 1850s (Berry, 1987:126–129). Beginning in the Jura Mountains (hence the name *Jurassic*) between France and Switzerland, he plotted the stratigraphic ranges of fossils from one locale to another, especially ammonites (a kind of cephalopod now represented by the modern octopus, squid, and pearly *Nautilus*; the class Cephalopoda in turn belongs to the phylum Mollusca, which includes clams and snails). Some species were short ranging (short lived), whereas others had intermediate or long ranges; the ranges were overlapping or concurrent (figure 2.3). From his plots, Oppel realized that there were distinctive aggregates of fossils from one locality to the next that he called zones, the upper and lower boundaries of which could be demarcated by the appearances of one or more species (he used appearances rather than disappearances of species because he felt that appearances were more likely to be unique events). Using this approach, Oppel zoned the entire Jurassic, and his zonal boundaries coincide with the upper and lower boundaries of the Jurassic system.

How are the zones matched (correlated) between sites? The stratigraphic range of a fossil species within a particular province is known as its teilzone (part or partial zone); the sum of the teilzones of a species everywhere within and between provinces constitutes its total range zone. Within a biogeographic province, the markers (zones) are related to those at other sites using the same species (because one is dealing with only a single province); the zonation (working hypothesis) is tested against fossil ranges at other sites and the zonation modified accordingly (the correlation web of Berry, 1987). This is true for *any* sort of biostratigraphic zonation.

What about correlation between provinces? Here one can use certain species that are more wide ranging geographically because they are more tolerant (eurytopic) of environmental change and therefore likely to occur in more than one province. One can also correlate between provinces by using the migration of a species (via dispersal of propagules—larvae, spores, seeds, or adults—by currents of water or wind) into or out of a province, as recorded by its appearance or disappearance in the rock record. Correlating zonations of different provinces is necessarily imprecise (eurytopic species tend to have longer geologic ranges than stenotopic ones and the appearance or extinction of a species in one province is unlikely to be exactly the same in another), but that is the best one can do. So a zonation within a particular province is unlikely to look exactly like the zonation within an-

FIGURE 2.3 Oppelian (overlapping, concurrent) zones for the Lower Silurian of Great Britain based on an extinct group of organic-walled planktic (floating) microfossils called graptolites. Similar zonations have been developed using other fossil taxa. Note that in many cases, the base of a zone coincides with the appearance of a new species. Names of zones are based on representative species. *From* Earth and Life Through Time *by S. M. Stanley. Copyright © 1989 by W. H. Freeman and Company. Used with permission.*

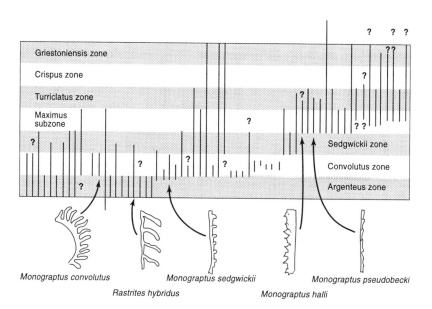

other province, even if the fossils used to zone the rocks belong to the same taxon (such as ammonites).

By establishing the total range zone of a species, one establishes what are called index fossils, which are useful for establishing age down to about the epoch (series) level. These species are sufficiently eurytopic to be widespread and reasonably abundant, but sufficiently stenotopic (narrowly tolerant) to undergo extinction fairly frequently and produce biostratigraphic markers. Hence, there is a tradeoff between being eurytopic (widespread) and stenotopic (with a short geologic range).

The concept of index fossils appears to have had its origins in Oppel's practice of naming each zone after a representative species (even though the species need not be confined to that zone; see chapter 5 of Eicher, 1976). Later paleontologists corrupted this practice and began to refer to zones by their names as if the nominal species were *confined* to the zone. Other workers, especially in the United States, used index fossils as if they were representative of particular formations. Obviously, for the reasons mentioned in our discussion of time and rock units, this was incorrect. It was also incorrect because many species, especially benthic (bottom-dwelling) ones, track environments (facies) and are called facies fossils. They may not be reliable time indicators because rock units (environments) cut across time lines. For the same reasons, peak or acme (ecostratigraphic) zones, in which one or more species (especially benthic ones) are abundant, have been disdained by many workers (but see Martin and Fletcher, 1995).

Nevertheless, both types of markers were successfully used by the petroleum industry before the advent of zonations based on planktic (floating) fossils because no markers were available other than benthic ones (a result of the ancient environments, which had been too close to shore) and because the value of planktic fossils was not realized until the late 1950s and 1960s. Planktic fossils are much preferred over benthic ones because they are not affected by the bottom environment. They are usually quite stenotopic, and tend to be widespread and common only in open ocean environments (outer shelf and beyond).

Based on the overlapping (concurrent) nature of the fossils' ranges, Oppel grouped zones into larger units called *Zonengruppen,* which are equivalent to stages. Despite the validity of Oppel's approach in Europe, it still has not been widely used in North America. Its initial application by Robert Kleinpell to the Miocene of California (finally published in 1933) was widely criticized, and the terms *stage* and *zone* were legislated out of existence at that time by committee (Berry, 1987:134–136), only to be recognized later.

In theory, stages may be grouped to produce series and series lumped to produce systems (to form a nested hierarchy), but that is not how it happened initially. Series were first formally recognized in about 1833 by Charles Lyell, who was working on rocks of the Cenozoic era located primarily in the Paris Basin. Instead of constructing a detailed zonation like that of Oppel, Lyell catalogued the number of still-living (extant) versus extinct species of molluscs because, as the age of the rocks increased, the number of living species represented in the rocks tended to decrease. Lyell avoided the facies problem by "averaging" it out (perhaps without really knowing that it existed; remember that the concept was recognized after Lyell published his work). He collected from many different types of rocks that had been deposited in different environments and lumped the geologic ranges together. Lyell recognized the epochs (series) Eocene (dawn recent), Miocene (middle recent), and Lower and Upper Pliocene (more recent); the latter would become the now-familiar Pleistocene (most recent), the last 10,000 years of which are often recognized as the Holocene (wholly recent). Other workers would soon supply the subdivisions Oligocene (few recent; recognized by von Beyrich in 1854 while working in Germany and Belgium) and Paleocene (ancient recent; split off from the lower Eocene by Schimper in 1874 based on fossil floras). So now we have a mnemonic for the Cenozoic: "Pigeon Egg Omelets Make People Puke Heartily."

Thus organic evolution had already been recognized by paleontologists from patterns in the fossil record before Darwin's work, although the mechanisms involved were (and still are, as discussed in chapter 5) a matter of dispute. Although Oppel's work was, in effect, based on organic evolution, he was apparently unaware of it, as his work was completed a year before publication of Charles Darwin's initial work (1859). Unlike his predecessor, Alcide d'Orbigny, who had also zoned the Jurassic using ammonites, he did not ascribe to divine intervention the succession of faunas (zones) in the Jurassic. The concept of stage was originally developed by d'Orbigny, who believed that each stage was sufficiently distinct to be correlatable around the world. D'Orbigny borrowed from another French paleontologist, Baron George Cuvier, who based his view of the history of life on a literal interpretation of the Bible. D'Orbigny's concept of stages resembled the layers of an onion, with the fauna of each stage being driven to extinction by a catastrophe, only to be replaced by the Creator (Berry,

1987:122–126). Although Darwin's theory stands as a triumph of the scientific method and rational thought, Darwin could never offer a fully adequate explanation for the mechanism of evolution. (Much the same fate befell Alfred Wegener and continental drift before the discovery of seafloor spreading ushered in the era of plate tectonics in the 1960s.)

Does the hierarchy of system, series, stage, age (period, epoch, age, chron) mean that the processes producing these units (patterns) are self-similar? Yes and no. A strict reductionist approach to a nested hierarchy does not apply here. Zones are typically recognizable only within a single biogeographic province, itself the smallest unit of another hierarchy (provinces, regions, and realms) used to describe the geographic distribution of organisms. Organisms are not uniformly distributed over the earth's surface, however, being concentrated within (or endemic to) certain areas because of differing tolerances to environmental factors (such as temperature, salinity, and light), pathways and barriers to migration (dispersal), and the vagaries of geologic history (such as uplift of mountains and transgressions of seas). With units of longer duration (such as stage and series), factors affecting the earth's biota over larger areas (regions and realms) presumably play an ever greater role. By the time we reach the period and system level, we are sometimes talking about the extinction of a large portion of the earth's biota and the subsequent reoccupation of vacant habitats by newly evolved taxa. This is not to say that environmental tolerance and dispersal ability are not important to survival during episodes of extinction, because eurytopic organisms tend to disperse widely and are able to withstand environmental shocks of greater magnitude than stenotopic ones (Jablonski, 1986a, 1986b, 1989; see also chapter 8). Moreover, contrary to the notion conveyed by the popular scientific press, ancient large-scale disturbances that caused species to go extinct or to migrate (and thus produce the biostratigraphic units seen in the rock record) more often than not seem to be unusual combinations of processes that may act at smaller spatial, and over shorter temporal, scales (see the discussion of holons in chapter 1). We return to this subject again in chapter 8.

REFINEMENT OF THE TIMESCALE

The Oppelian approach is not widely used today. Instead, with the advent of planktic zonations in the 1950s and 1960s based on foraminifera, calcareous nannofossils (such as coccolithophorids), and siliceous plankton (radiolarians and diatoms, among others), the use of first and last appearance datums (FADs and LADs) of certain marker species has become common (see also Berry, 1987, and Boggs, 1987, chapter 18; if FADs and LADS are in doubt, they are usually denoted as lowest and highest occurrences [LOs and HOs], respectively). Planktic fora-minifera have served as the basis for subdivision of Cenozoic rocks into a series of numbered zones (P1–P22) for the Paleogene (Paleocene through Oligocene) and Neogene or Miocene to Recent (N1–N23). A similar zonation has been erected using calcareous nannoplankton. Cenozoic sediments are quite prominent not only on land, but also in the deep sea, and the Deep Sea Drilling Program or DSDP (and its successor, the Ocean Drilling Program or ODP) and the strong interest of petroleum companies have driven subsequent refinements of these zonations. Moreover, because of the need for more precise correlation, other types of markers, such as magnetic reversals (which are synchro-nous worldwide and so provide anchor points if they can be dated), oxygen isotope datums (which reflect changes in ice volume and are most distinctive in the last half of the Pleistocene), and most recently strontium isotope datums (more on these methods in later chapters) have been integrated into the biostratigraphic schemes to produce a highly calibrated chronostratigraphic or temporal scale for the Ceno-zoic. This scale is constantly being refined and recalibrated based on new results from ODP cruises and work on outcrops.

Two important points must be made here. FADs and LADs are just as likely to be diachronous as other datums. The first or last appear-ance of a marker in one ocean basin or biogeographic province is not necessarily the same in another, hence the need for calibration using independently derived datums from other methods (such as paleomag-netics). However, it would be a mistake to think that these other disci-plines are superior to biostratigraphic approaches. Magnetic reversals are basically alike (an interval is either reversed or normal), although the exact boundary between normal and reversed intervals is not al-ways as clear-cut as some workers would have you believe (see the signal/noise ratios for paleomagnetic curves plotted in DSDP and ODP volumes). Oxygen isotope stages are also similar from one stage to another (ice advances or retreats), although each stage (especially in the last half of the Pleistocene) tends to have a particular character that sets it apart from others to a trained eye. Basically, then, these datums repeat themselves through time, just as rocks do, and it is for these sorts of datums that the biostratigraphic datums tell us where we are in the geologic column. The refined scale, then, is dependent on both types of datums, not one or the other.

THE FRACTAL NATURE OF THE GEOLOGIC RECORD

Are other aspects of the stratigraphic record potentially self-similar (in a statistical sense)? Perhaps. Let's return to the distinction between time and time–stratigraphic units. Say that a particular spot on the earth's surface started out with an uninterrupted sediment accumulation rate of 10 cm/1000 yr (ka); this is not an unusual rate for, say, river deltas, which are sites of significant sediment influx (Enos, 1991; see also Schindel, 1980). Then over the course of the Phanerozoic (the last 600 million years or 600 mega-annums [Ma]), about 60 km of sediment would have accumulated. This thickness is three times that of the average thickness of continental crust! Clearly, much of the sediment must have bypassed the site of deposition or was eroded. In fact, a general pattern emerges that has been noted by a number of workers over the years (such as Sadler and Strauss, 1990): A log–log plot of sedimentation rate and the period of observation over which the rate was calculated results in a line with a negative slope (downward to the right). This means that sedimentation rates are fast when measured over short time spans, but considerably slower when measured over longer intervals. More and longer intervals of erosion and nondeposition are included as the duration of observation increases. Moreover, as the slope of the line increases, the relative duration of the stratigraphic gaps (G) increases (figure 2.4).

The time represented by the missing sediment is called a hiatus (sometimes on the order of hundreds of thousands to millions of years or more) and the surface of erosion or nondeposition is called an unconformity. It was a particular type of unconformity, called an angular unconformity—in which stratified rocks are first tilted from the horizonal, then eroded and eventually covered with more flat-lying sediments—that gave James Hutton, in 1788, at Siccar Point, Scotland, his inklings about the great age of the earth and that led to the principle of uniformitarianism. Unconformities are only the most obvious expression of erosion or nondeposition of sediment.

Unconformities are the basis for the discipline of seismic stratigraphy (later dubbed sequence stratigraphy), which caused a revolution in the science of stratigraphy, especially within the petroleum industry, which immediately seized upon it when it was first published in 1977.

An outgrowth of Lawrence Sloss's work on the North American craton (continental interior) in the 1950s and 1960s (Sloss, 1963), seismic stratigraphy was developed by Peter Vail, Robert Mitchum, and colleagues at Exxon Production Research in Houston (Vail et al., 1977) in an attempt to better predict the location of oil and gas reservoir sands, and thereby save on exploration costs. (A widely circulated story is that Exxon management allowed publication of Vail et al.'s work because it saw no practical use for it.) The basic unit of seismic stratigraphy is the depositional sequence, which consists of rock units bounded above and below by unconformities. Depositional sequences are thought to be cyclic in nature (related to sea-level fall and rise) and to represent durations of several hundred thousand to a few million years (Wilgus et al., 1988). The exact cause of sea-level cycles of intermediate duration is not known because increased seafloor spreading rates (and increased midocean ridge volume) are too slow, and glacial advances and retreats too fast, to account for them, but this has not

FIGURE 2.4 Log–log plot of sediment accumulation rate versus time span of observation (analogous to ℓ in chapter 1) for different gap (G) values. *From R. E. Plotnick, Journal of Geology, v. 94, p. 886. © 1986 University of Chicago. Reprinted by permission.*

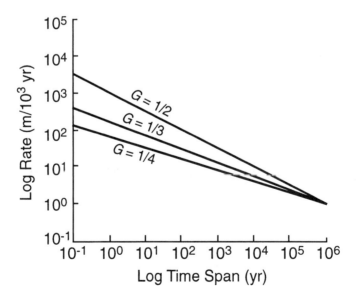

prevented numerous workers from using the concepts anyway (see chapter 6 for further discussion).

At the base of each depositional sequence is an unconformity (sequence boundary) formed through erosion related to rapid regression (fall) of sea level (figure 2.5; see Holland, 1995, for a succinct, jargon-free discussion of sequence stratigraphy). As sea level continues to fall, river deltas prograde across the shelf, and if sea level falls enough, eventually the deltas dump sediment at or beyond the shelf edge onto the continental slope to form a type of deposit called a lowstand systems tract (or LST). As sea level begins to rise and deltas retreat shoreward, a transgressive systems tract (TST) begins to form, and as the rate of sea level rise slows and sea level approaches its maximum height, a highstand systems tract (HST), indicative of net upward shallowing, forms. The maximum height in sea level that represents the transition from TST to HST is called a maximum flooding surface (MFS). The MFS is commonly associated with entrapment of sediment close to shore in estuaries, bays, and lagoons, so that across the continental shelf and slope only fairly thin units, called condensed sections, are formed. Condensed sections are highly fossiliferous (the other systems tracts may also be fossiliferous), especially offshore, and consist of deeper-water benthic fauna and diverse planktic microfossils that have settled out of the water column and may serve as biostratigraphic markers.

At the other extreme from unconformities are diastems, which are gaps in the stratigraphic record so short that they are virtually undetectable. Diastems are caused by very short-term changes in sedimentary regimes. Sedimentation within a particular environment is not uniform and continuous throughout; storms, for example, may redistribute previously deposited sediment but leave no record of themselves. In between unconformities and diastems are stratigraphic gaps and hiatuses of all magnitudes and durations. The distinction between sedimentation and sediment accumulation (and the reason for time and time–stratigraphic units) should now be clear.

Based on Barrell's (1917) famous diagram (figure 2.6), one might suspect that the distribution of stratigraphic hiatuses is fractal. Embedded within a depositional sequence are smaller-scale "paraquences," which undergo net shallowing or deepening trends (Wilgus et al., 1988; Holland, 1995). Using log–log plots of sediment accumulation rate and duration of observation, Plotnick (1986) developed a fractal model ("an intellectual construct for organizing experience"; Allen and Starr,

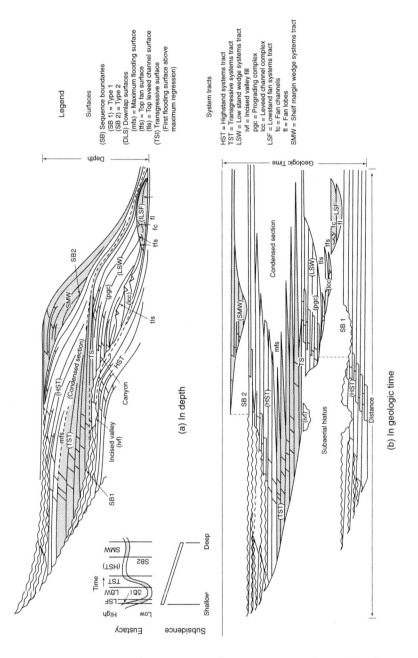

FIGURE 2.5 Depositional sequence and systems tracts for a siliciclastic setting. (a) Geometry of systems tracts. (b) Same as in (a) but in relation to geologic time. *From B. U. Haq, J. Hardenbol, and P. R. Vail. 1988. Mesozoic and Cenozoic chronostratigraphy and cycles of sea-level change. In C. K. Wilgus et al., eds.,* Sea-Level Changes: An Integrated Approach. *Tulsa, Oklahoma. Society of Economic Paleontologists and Mineralogists Special Publication No. 42. 71–108. Reprinted with permission of SEPM (Society for Sedimentary Geology).*

1982) for hiatus distribution based on the concept of a Cantor bar, in which the size of the stratigraphic gaps (G) is allowed to vary (figure 2.7; for the sake of simplicity, I use the word *model* as a synonym for *working hypothesis*). The rocks and the gaps are the curds (clumps) and whey (milky liquid), respectively, of Mandelbrot's (1983) terminology, with periods of sedimentation being clustered in time.

The Cantor bar model provides a theoretical distribution of deposition versus hiatuses for a particular environmental setting *before* one studies it (it is also consistent with compilations of sedimentation rates in different environments published by other workers; Plotnick, 1986; also see Sadler, 1981; Sadler and Strauss, 1990; and McKinney, 1991). Thus the suitability of a particular depositional setting for, say,

FIGURE 2.6 The sedimentary rock record (left) and the amount of time recorded in rocks and depositional gaps (top) for different rates of sediment accumulation (A, B, and C). *From C. O. Dunbar and J. Rogers,* Principles of Stratigraphy, *Wiley, New York, 1957; after Barrell, 1917. Reprinted with permission of J. Rodgers, the trustee of the estate of Carl Dunbar, and the Geological Society of America.*

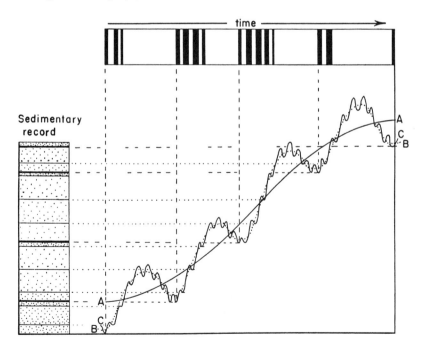

paleobiological or stratigraphic studies can be assessed beforehand, thereby saving time and money.

The slopes of the lines of the log–log rate versus time plot (figure 2.8) are calculated from

$$m = \frac{-\log(1 - G)}{\log[(1 - G)/2]}. \tag{2.1}$$

Let T_n (S_n) and $T_{n + 1}$ $(S_{n + 1})$ be successive time (T) intervals of observation represented by sediment (S) thickness. Sedimentation rates (R) are then calculated by dividing sediment thickness by time. Assume $R_{n + 1} > R_n$ and let $G = 1/2$. Then for a particular interval,

$$T_{n+1} = \frac{T_n(1 - G)}{2} \tag{2.2}$$

and

$$S_{n+1} = \frac{S_n}{2}. \tag{2.3}$$

FIGURE 2.7 Cantor bar generated with different gap (G) sizes. Vertical axis represents time. For each gap size, the amount of time represented by rock decreases with each iteration. Compare with figure 2.6. *From R. E. Plotnick,* Journal of Geology, *v. 94, p. 886. © 1986 University of Chicago. Reprinted by permission.*

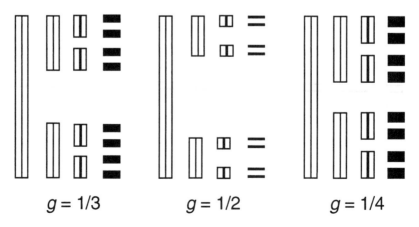

$g = 1/3 \qquad g = 1/2 \qquad g = 1/4$

FIGURE 2.8 Relationship between sediment accumulation rate and time span of observation for a gap (G) value of ⅓. *From R. E. Plotnick, Journal of Geology, v. 94, p. 886. © 1986 University of Chicago. Reprinted by permission.*

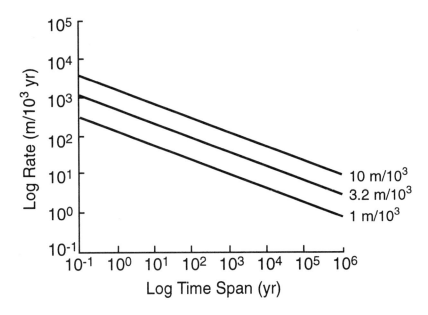

Because slope (m) = rise/run, on the log–log plot of rate (R) versus time (T)

$$m = \frac{\log R_{n+1} - \log R_n}{\log T_{n+1} - \log T_n}. \qquad (2.4)$$

Substituting from equations 2.2 and 2.3,

$$m = \frac{\log[(S_n/2)/T_n(1 - G)/2)] - \log(S_n/T_n)}{\log(T_n(1 - G)/2) - \log T_n}. \qquad (2.5)$$

Simplifying and canceling terms results in equation 2.1. Based on the difference in completeness between short and long time spans, an estimate of the completeness of the section for a particular span of time can be made from

$$\text{Completeness} = (t/T)^{-m}, \qquad (2.6)$$

where t = desired short time span, T = time span of the whole section, and m = slope of the line (equation 2.1).

The Cantor bar model goes beyond estimating the completeness of section to estimating the number of hiatuses and their length for the chosen period of observation (t). Rearranging equation 2.1,

$$\log(1 - G) = (m/m + 1)\log 2. \qquad (2.7)$$

For example, from equation 2.7, for t = 1000 yr, Plotnick (1986) estimated a 4-million-year Eocene section in Wyoming to be 11.6 percent complete $(m = -0.26)$. Assuming 1000-yr intervals of deposition, 88.4 percent of the time is represented by hiatuses. For the same section, $G = 0.22$, which was then used to calculate iteratively that the Eocene section consisted of 256 periods of deposition, each 2141 yr long, separated by 255 hiatuses ranging in length from 1208 to 880,000 years. Fortunately, things are better for deep-sea records, although they too are often incomplete (see Plotnik, 1986, for examples).

There is a tradeoff, then, between rates of sedimentation and continuity of sedimentation. If one wanted to study, say, fossil assemblages that are undisturbed or of short temporal duration, one might pick a deltaic environment because sedimentation can be quite fast (up to 200 m/ka; Enos, 1991). As discussed in chapters 3 and 4, rates of sedimentation have important consequences for rates of sediment mixing by organisms (bioturbation) and disruption or loss of high-resolution information in the fossil record. The disadvantage in picking a deltaic environment is that the sedimentary column for such an environment tends to occur in "packages" of rapidly deposited sediment, separated by unconformities, so that the record can be very discontinuous (Schindel, 1980, 1982). The deep-sea record, on the hand, is characterized by slow sedimentation (about 2 cm/ka average), but sedimentation consists of a constant rain of pelagic (primarily microfossils) or hemipelagic sediment (microfossils deposited with terrigenous land-derived sediment).

Although the distribution of hiatuses and sedimentation rates is fractal (self-similar), it appears to be hierarchical, which suggests that different processes act at different spatio-temporal scales (Plotnick, 1986); thus there appears to be a nested self-similarity (McKinney, 1991). In a delta, short-term changes in sedimentation (and hiatus frequency) are controlled by such factors as lobe switching, whereas longer-term sedimentation rates and hiatuses reflect such processes as sea-level change. Moreover, sea-level change, sediment availability (erosion rates), and "accommodation space" (a place to put the sediment) exert top-down control on the position and behavior of delta lobes (Plotnick, 1986; McKinney, 1991). Sea-level change itself reflects a variety of processes acting at different spatio-temporal scales (chapter 6).

THE STRATIGRAPHIC DISTRIBUTION OF FOSSILS

Returning to the previous discussion of the stratigraphic distribution of fossils and zonation, what effect does the distribution of rocks and unconformities have on the distribution of fossils? Not all systems tracts, for example, are present in any one sequence; their occurrence depends in part on sediment input (such as delta-lobe switching) and location with respect to the continental margin. Sequences located on the outer continental shelf are usually complete initially, but may lose the HST during the erosive phase of the subsequent LST. In continental interiors (well up on the craton), only the TST and HST are usually found because this area represents the furthest landward penetration of sea level. The missing sediments are represented by the sequence boundary (Holland, 1995).

This pattern is often obvious in range charts of microfossil distributions produced by micropaleontologists (figure 2.9). High sea level is represented by well-developed fossil assemblages (numerous thick lines indicating high diversity and abundance), whereas barren intervals indicate very shallow waters that are normally represented by low-diversity faunas and high sediment influx (which dilutes fossil assemblages). The abrupt contact below the two types of intervals represents an unconformity (rather than a fault) if it can be traced over wide distances. In this example, sand is associated with the sequence boundaries and may represent deposition during an LST or subsequent reworking (basal lag sands) during the subsequent transgression (TST). The only way to really tell is to determine the facies relationships from seismic sections *integrated* with the fossil data, which give a true record of the depositional environment (see Armentrout, 1991, 1996).

Most recently, Holland (1995) has developed a model that simulates the stratigraphic distribution of fossils in relation to systems tracts. First and last occurrences of fossils (which may represent evolutionary appearances and extinctions within a province) cluster at sequence boundaries and flooding surfaces of TSTs (figure 2.10). This is because the true upper or lowermost occurrences of fossils are lost through erosion (sequence boundaries) or are not present because the fossils themselves may track environmental (sea level) change (just as rocks do). When Holland added gradients in diversity (lowest nearshore, highest offshore), taphonomy (preservation), and environmental tolerance that may occur across continental shelves, the intensity, but not the position of the stratigraphic peaks varied. Importantly, Holland (1995) went beyond the computer and tested the model (hypothesis)

FIGURE 2.9 Fossil range chart for petroleum well drilled in the Gulf of Mexico. Foraminifera are unicellular shelled protozoans, whereas nannofossils secrete platelets of CaCO$_3$. Thickness of each line indicates relative abundance of a species, which was estimated visually by oil company paleontologists. Lithological (sand, cement) descriptions based on visual estimates and well-log information. Sequence boundaries (based on graphic correlation) are also shown. *Modified from Martin and Fletcher, 1995; original data from Unocal.*

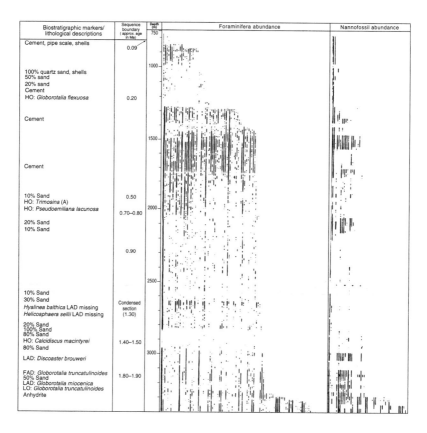

FIGURE 2.10 Numbers of first and last occurrences of fossil species through two depositional sequences in a single stratigraphic section deposited in a shallow seaway ("upramp" of the modern continental shelf). LST = lowstand systems tract, TST = transgressive systems tract, and HST = highstand systems tract. In this particular example, no deposition occurs in the LST because the setting is too far onto the craton, and the time represented by the LST is contained in the sequence boundary (located between LSTs and TSTs). Any peak above the dashed line is statistically significant. Significant numbers of first occurrences are associated with TSTs, whereas significant last occurrences tend to occur below sequence boundaries. See also figures 2.5 and 2.9. *Modified from Holland, 1995.*

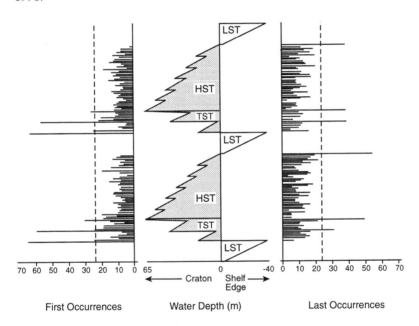

against real-world data from the Late Ordovician (Cincinnatian Series) and found that his modeled occurrences are present in the fossil record.

Even if the stratigraphic record is complete, as we approach some biotic event (say a large extinction) or other boundary, and even if the event is presumably sudden (a result of catastrophic causes), only the most abundant species persist to the boundary. Less abundant species tend to disappear from the record somewhere below the boundary, never to reappear (at least in a particular section) because they are uncommon enough that they are unlikely to be found. This sampling effect has been named the Signor–Lipps effect, after its describers (Signor and Lipps, 1982; see also Dingus, 1984). Other species may appear to die out before or during an extinction, only to be resurrected in the rock record after (above) the event has occurred (Lazarus taxa), whereas others that do become extinct may be replaced ("impersonated") by superficially similar, but quite different, groups called Elvis taxa (Erwin, 1993).

Meldahl (1990; see also Flessa, 1990) simulated sudden (catastrophic), stepwise, and gradual extinctions (patterns that have been suggested by other workers; see Raup, 1989, and Pospichal, 1994) based on abundances of molluscs (mainly clams and snails) sampled at 1-cm intervals in 8 cores taken through modern tidal flat sediments of Bahia la Choya, Mexico (northern Gulf of California). For a catastrophic extinction, he assumed that the extinction event occurred at the top of the cores. For gradual and steplike extinctions he used the distribution of molluscs in the cores, and made certain species "extinct" by eliminating all biostratigraphic occurrences above certain levels in the cores. For gradual extinction, he selected species one by one for extinction at successive 1-cm intervals, beginning at 44 cm depth. For stepwise extinction, he chose 15 species at random at the 40-cm level in cores, 15 other species for extinction at the 20-cm level, and the remaining 15 species at the core tops. He ran each simulation five times, and found that "extinction" of a species was more likely to be recorded as sampling intensity increased (figure 2.11), and that species that occurred in less than 15 percent of his samples did not have their uppermost occurrences accurately recorded (this amounted to 71 percent of the 45 species sampled). Each type of extinction produced a decline in diversity (number of species) that differed in pattern as the extinction boundary was approached. Sudden extinction produced a diversity decline that intensified as the boundary was approached, whereas gradual extinction produced a constant decline, and stepwise extinction resulted in a stepwise decline, although this pattern was difficult to distinguish from the gradual one.

Thus the observed stratigraphic range of a species almost always underestimates its true range. This relationship can be expressed in the form of confidence intervals (CIs),

$$\alpha = (1 - C_i)^{-1/(H-1)} - 1 \qquad (2.8)$$

FIGURE 2.11 Comparison of stratigraphic ranges of mollusc species in a single core versus ranges compiled from eight cores taken through modern tidal flat sediments in the northern Gulf of California. As the number of samples increases, error in stratigraphic ranges decreases and uppermost occurrences of fossil species shift upward toward hypothetical extinction boundary (core top). New species are also found as sampling continues. *From K. H. Meldahl, 1990. Sampling, species abundance, and the stratigraphic signature of mass extinction: a test using Holocene tidal flat molluscs. Geology 18:890–893. Reprinted with permission of the Geological Society of America.*

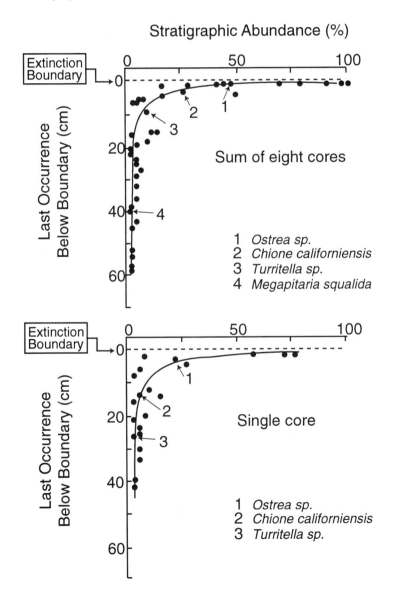

(Marshall, 1990; equation 2.8 is derived by Strauss and Sadler, 1989), where α = the confidence interval expressed as a fraction of the known stratigraphic range, C_i = desired confidence level, and H is the number of known beds from which the species has been sampled. For example, if $H = 10$, then for a C_i of 95 percent (that is, you are certain that in 95 percent of the beds sampled the species will occur within the calculated range), $\alpha = 0.39$. If the oldest known bed is 60 Ma and the youngest 50 Ma (range = 10 million years), then the appearance of the fossil species probably occurred (with 95 percent confidence) between 60 and 63.9 Ma and its disappearance probably occurred between 50 and 46.1 Ma. α values at a C_i of 0.5 are much smaller than those for a C_i of 0.95. If both endpoints are to be estimated simultaneously, a slightly different version of equation 2.8 is used.

This approach may be used on both local sections and composite sections developed from a number of different local sections. The main assumptions using the method are that fossil horizons are randomly distributed (meaning that sedimentation rates and fossilization potential remained constant through time), the intensity (efficiency) of collection was uniform throughout each section, and fossilization events are statistically independent of one another. The assumptions are more easily fulfilled for local sections than for composite ones because of variation in sampling by investigators working on different sections or variation in preservation from outcrop to outcrop. Marshall (1990) explores these limitations more thoroughly.

Marshall (1995) calculated 50 percent confidence intervals to test the null hypothesis of a sudden extinction recorded by ammonite range data near the Cretaceous/Tertiary (K/T) boundary in Antarctica. He used 50 percent intervals because for a single taxon there is a 50/50 chance that the true extinction of a species lies above or below the top of the 50 percent confidence interval for the species. Thus if more than one species went extinct simultaneously, then on average half of the upper confidence intervals calculated for the species assemblage should occur above the level of extinction and half below it. Based on this approach, he could not reject the null hypothesis of a sudden extinction (notice that I did not say that he *accepted* the null hypothesis). He calculated that the most likely position of the K/T boundary, assuming that the ammonites became extinct at the boundary, was 7 to 11 m *above* the last recorded ammonite, but that the K/T event could have occurred anywhere between 0 and 20 m above the last ammonite. An iridium layer (which is presumably indicative of an impact at the end of the Cretaceous; see chapter 8) also occurred within the calculated confidence intervals. The data and calculated ranges are also consistent with gradual extinction and more stepped extinctions as long as the extinctions did not occur over stratigraphic intervals greater than 20 m. He concluded that the only way to distinguish between sudden and gradual extinction scenarios is to conduct saturation collecting near the K/T boundary.

GRAPHIC CORRELATION

Is there a way to tell where stratigraphic gaps actually occur or whether a particular biostratigraphic (fossil) marker occurs too high or too low in a particular outcrop or core? Yes, there is. The method has been with us for about 30 years, but it has not been until the last 5 to 10 years, with the widespread availability of powerful personal computers and the availability of privately developed and commercial software (see Hood, 1986), that the technique of graphic correlation has begun to flourish.

Graphic correlation is a technique that can be used to test the synchroneity of traditional (biostratigraphic) and nontraditional (oxygen isotope stage, paleomagnetic, well-log, microtektite and volcanic ash layers, ecostratigraphic markers) and to detect missing (and expanded) sections (see Edwards, 1989a; Glass and Hazel, 1990; Martin et al., 1990, 1993; Martin and Fletcher, 1995). The method incorporates sections (cores, wells, outcrops, or any combination) into a composite section, against which individual sections are later replotted (Shaw plots, named for Alan Shaw, an Amoco geologist who developed the technique to study oil wells in the Rocky Mountain overthrust belt). Datums that plot on or close to a line of correlation (LOC), which is normally fitted based on investigator experience, are considered coeval (synchronous). *Any* length of section may be used as long as suitable stratigraphic markers are available for its subdivision.

The LOC indicates changes in sediment accumulation rate. An LOC that is at approximately a 45° angle from either the vertical or horizontal indicates that the sediment accumulation rates of the two sections are equal (figure 2.12). Normally, however, the LOC consists of a series of segments ("doglegs"), the angles of which differ from 45°, indicating that the two sections differ in their respective accumulation rates. For example, in figure 2.12, high-angle (from the horizontal) LOC segments indicate that the accumulation rate of the section plotted on the vertical (y) axis is higher than that of the section plotted on the horizontal (x) axis. Conversely, low-angle (from the horizontal, in

FIGURE 2.12 Changes in sediment accumulation rate are reflected by changes in slope of the line of correlation in plots of stratigraphic sections against one another. Line of correlation is normally fitted based on investigator experience using different stratigraphic datums. *Modified from Phillips, 1986.*

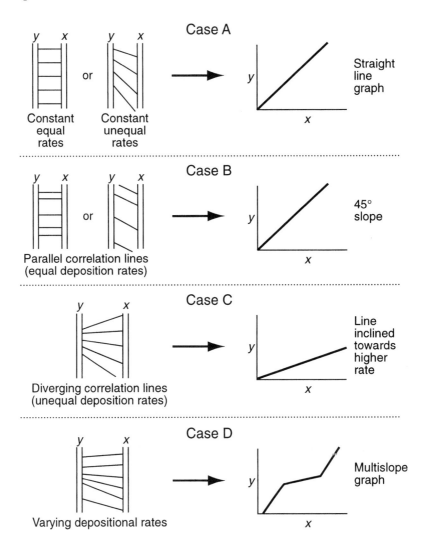

this case) line segments indicate that the net sedimentation rate of the section plotted on the x-axis is higher than that of the section plotted on the y-axis. Sudden changes in slope of the LOC may indicate unconformities (or sequence boundaries; see Martin et al., 1993), especially if they coincide with truncated ranges of stratigraphic markers. Stratigraphic markers that lie well off (above or below) the LOC are suspect because they may be reworked (by erosion or possibly bioturbation) or represent truncated ranges or shifting environments (facies).

Graphic correlation is initiated by choosing one section as the standard reference section (SRS). Normally, this is done by making preliminary plots of sections against each other, which are then evaluated based on investigator experience for continuity and length of section, and stratigraphic resolution (number and spacing of stratigraphic markers). It would probably also be helpful to evaluate confidence intervals of appropriate taxa before choosing an SRS (Strauss and Sadler, 1989). Martin and Fletcher (1995) review some of the pitfalls in choosing an SRS.

Once the SRS has been chosen, other sections are plotted against the SRS to construct a composite section by a process called range extension (see Shaw, 1964; Miller, 1977; Edwards, 1989b; and Mann et al., 1995, for reviews). Range extension maximizes the stratigraphic range of markers. The composite section thus assimilates datums from all sections and displays their maximum stratigraphic ranges in a single, hypothetical stratigraphic column. It is like taking all the individual sections of a fence diagram and combining them into a single stratigraphic column. The composite is normally subdivided into equal increments called composite standard units that are equivalent to time–stratigraphic units (rocks bounded above and below by time). Range extension is repeated for each section because the first and subsequent sections were not originally plotted against a composite based on all sections. Once completed, the sections are replotted until all datums have stabilized in position. Graphic correlation, then, is a process by which one erects a working hypothesis (SRS) that is tested and retested by plotting and replotting sections against one another to produce a well-substantiated hypothesis or theory (composite section), much like a correlation web. In doing so, the best approximation of the geologic ranges of taxa are determined, despite the vagaries of preservation and sampling (see chapters 3, 4, and 5).

Using this technique, stratigraphic gaps may be detected that are not apparent to the unaided eye. Martin and Fletcher (1995) used the technique to demonstrate missing and condensed sections in what *appeared* to be continuous section cored through the Plio-Pleistocene of the Gulf of Mexico and to construct a sea-level curve that closely matched those published by other workers using extensive (and expensive) integrated seismic, well-log, and paleontologic data bases (see Armentrout, 1991, 1996). Based on Shaw plots, MacLeod and Keller (1991) and MacLeod (1995) suggested that a number of sections previously analyzed for the Cretaceous/Tertiary (K/T) extinctions and presumed to be complete are actually represented by missing section.

Obviously, unconformities are common throughout the geologic column, and diastems even more so. But there is one process that is even more common and insidious: bioturbation.

3

Random Walks in Muck

$$\int_{-\infty}^{t} N(t')Q(t - t')\, dt'. \quad -Allen\ and\ Starr$$

No, the quotation is not in Greek. It is actually a filter, where N = signal (the units are unimportant for our purposes) and Q = a weighting function, which specifies the weight to attach to the signal at past times to produce a lag $(t - t')$.

Any signal that is incorporated into the rock (fossil) record must first pass through one or more such low-pass filters, in which high-frequency events are damped or removed and lower-frequency events preserved. A variety of filters exist, but the one we review at length in this chapter is *bioturbation* (mixing of sediment by organisms). Most earth scientists have studiously avoided this process over the years, but if we wish to extract and assess high-resolution data in the rock record, we must ultimately confront this process head-on. Bioturbation is a major impediment to bridging the gap between ecological (short-term) and geological or evolutionary (long-term) processes (other impediments are discussed in chapter 4). On the other hand, we may view it more favorably as erasing high-frequency noise and leaving behind evidence of longer-term patterns and their processes (Wilson, 1988; McKinney, 1991).

A ONE-DIMENSIONAL RANDOM WALK

Among the most popular sediment mixing (bioturbation) models are diffusion models (see Matisoff, 1982; Cutler, 1993; and Martin, 1993 for reviews). Diffusion models assume that sediment mixing can be described mathematically as a random diffusion process: An eddy (particle) biodiffusion coefficient accounts for the redistribution of sediment particles by large numbers of organisms over a large number of individual transport events. Taken collectively, small transport events may move particles over much larger distances. In one of the most frequently cited diffusion models, Guinasso and Schink (1975) modeled concentration profiles of microtektites in deep-sea sediments using the equation

$$\frac{D\partial^2 c}{\partial x^2} - \frac{v\partial c}{\partial x} - \lambda c = 0, \tag{3.1}$$

where
c = concentration of the tracer (cm^{-3}),
D = eddy biodiffusion coefficient $(cm^2\ ka^{-1})$,
v = sedimentation rate $(cm\ ka^{-1})$,
λ = radioactive decay coefficient (ka^{-1}), in the case of radioactive tracers, and
x = depth (increasing downward in cm).

What this equation says is that the rate of change of tracer activity at depth x owing to mixing minus burial of tracer and tracer decay equals 0; in other words, $\partial c/\partial t = 0$, and the system is assumed to be in steady-state (i.e., what goes in is what comes out). Biodiffusion coefficients vary by about six orders of magnitude and tend to decrease from shallow water $(D \approx 10^{-6}\ cm^2/sec)$ to the deep sea $(D \approx 10^{-8}\ cm^2/sec;$ Matisoff, 1982). Where exactly did this equation come from and what is the meaning of a biodiffusion coefficient?

Much of the development of equation 3.1 can be found in chapters 1, 2, and 4 of Berg (1993). At absolute temperature T, a particle exhibiting Brownian movement has an average kinetic energy of $kT/2$ (k = Boltzmann's constant), irrespective of particle size (up to and including those seen under a microscope). This kinetic energy may also be expressed as $<Mv^2/2>$, where M = mass, v = velocity, and $<>$

denotes an average over time or the average of a group (ensemble) of similar particles. From this relationship, the mean-square velocity and root-mean-square velocity, respectively, can be calculated:

$$<v^2> = kT/M \qquad (3.2)$$

and

$$<v^2>^{1/2} = (kT/M)^{1/2}. \qquad (3.3)$$

Using these equations, Berg (1993) calculated a mean-square velocity of a small protein particle (lysozyme) of 1.3×10^3 cm/sec (about 47 km/hr). In other words, an unimpeded molecule of lysozyme would fly across a room in no more than a few seconds. Of course other particles impede the progress of this molecule, so its velocity is much slower. In fact, it encounters other particles so often that it undergoes a random walk (diffusion) unless an external force (drift; in our case, burial) is applied.

Berg (1993) described a simplified model of diffusion, the equations of which eventually lead back to equation 3.1. Assume the following conditions: (1) an ensemble N of particles start movement at time $t = 0$ and position $x = 0$ and move along only one axis (the x-axis) for the sake of simplicity; (2) each particle moves to the right or left once every τ seconds at velocity $\pm v$ (figure 3.1); therefore, each particle moves a distance (or step length) $\delta = \pm v\tau$; (3) each time a particle moves, it has a 50/50 chance of moving to the right or left, and each movement (step) of a particle is statistically independent of the previous and subsequent steps; that's why the model is a called a random walk; (4) each particle does not interact with other particles (of course, this is not true, but it simplifies the model greatly; we discuss criticisms of the diffusion-based analogy of bioturbation later).

Now see what happens to a single particle based on these rules. Let $x_i(n)$ be the position along the x-axis of the ith particle after the nth step. The position of the particle after the nth step differs from its position at the $(n - 1)$ step by $\pm \delta$, or

$$x_i(n) = x_i(n - 1) \pm \delta. \qquad (3.4)$$

The average movement of the ensemble of particles is then

$$<x(n)> = \frac{1}{N}\sum_{i=1}^{N} x_i(n). \tag{3.5}$$

Based on equation 3.4, equation 3.5 can be expressed as

$$<x(n)> = \frac{1}{N}\sum_{i=1}^{N} [x_i(n-1) \pm \delta]. \tag{3.6}$$

The term δ in equation 3.6 approximates 0 because each particle has a 50/50 chance of moving to the right or left at each step. Therefore,

$$<x(n)> = \frac{1}{N}\sum_{i=1}^{N} x_i(n-1), \tag{3.7}$$

which is the average particle displacement at step $n - 1$ and is expressed as $<x(n-1)>$. Based on this set of assumptions, on average,

FIGURE 3.1 Particles undergoing a one-dimensional random walk starting at the origin $(x = 0)$ and time $t = 0$, and moving in steps of length δ. The area under the curve represents the probability of finding a particle at different points along the x-axis at times $t = 1$, 4, and 16. The standard deviations (root-mean-square widths) of the distributions increase with $\sqrt{t}$, whereas peak heights decrease with $\sqrt{t}$. Compare with figure 3.5. *Redrawn from Berg, 1993.*

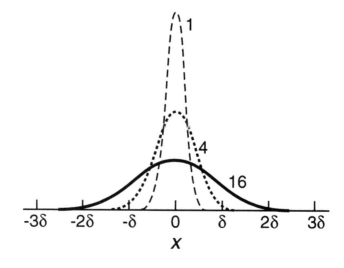

the particles move very little. Nevertheless, the particles in our example do spread away from the origin ($x = 0$).

How much do the particles spread away from their original position (figure 3.1)? This can be determined using the mean-square of the average particle displacement. The square of the displacement itself is

$$x_i^2(n) = x_i^2(n - 1) \pm 2\delta x_i(n - 1) + \delta^2. \qquad (3.8)$$

(Look again at the right-hand side of equation 3.4 and square it using the quadratic equation $a^2 + 2ab + b^2$). The *mean*-square displacement is then (cf. equation 3.6)

$$<x^2(n)> = \frac{1}{N}\sum_{i=1}^{N}[x_i^2(n - 1) \pm 2\delta x_i(n - 1) + \delta^2]. \qquad (3.9)$$

Because $\delta \approx 0$, equation 3.9 simplifies to

$$<x^2(n)> = \frac{1}{N}\sum_{i=1}^{N}x_i^2(n) \qquad (3.10)$$

(cf. equation 3.7), which is the same as $<x^2(n - 1)> + \delta^2$ (cf. equation 3.4). Based on equation 3.10, because $x_i(0) = 0$, $<x^2(0)> = 0$; thus, for $n = 1, 2, 3, \ldots$, $<x^2(1)> = \delta^2$, $<x^2(2)> = 2\delta^2$, $<x^2(3)> = 3\delta^2, \ldots$, $<x^2(n)> = n\delta^2$. Therefore, average particle dispersion (mean-square displacement) increases with step number (n), and mean-square displacement thus increases with time because $t = n\tau$. The actual amount of spreading (the *root*-mean-square of the displacement) is thus proportional to $\sqrt{t}$. To sum up, particle spreading increases with the square root of time, not time itself (figure 3.1).

We now return to the meaning of the diffusion coefficient (D). Because $n = t/\tau$, $x(n) = x(t/\tau)$, and

$$<x^2(t)> = (t/\tau)\delta^2 = (\delta^2/\tau)t \qquad (3.11)$$

(see the discussion in the last paragraph regarding equation 3.10). If we define $D = \delta^2/2\tau$, then

$$<x(t)^2> = (2Dt) \qquad (3.12)$$

and

$$<x(t)^2>^{1/2} = (2Dt)^{1/2}. \qquad (3.13)$$

Remember that the distance a particle moves is proportional to $\sqrt{t}$. Therefore, in order for a particle to move twice as far, it takes four times as long, and so on. Because particles usually move only short distances, they tend to return to the same point many times before finally leaving an area (figure 3.2). Particles wander to new areas at random (statistically independent of previous positions; that is, they walk randomly), and they tend to "explore" a new area fully before moving on to another region. The track of the particle does not necessarily fill space uniformly.

Imagine figure 3.2 as if one were either looking down on the sediment–water interface or looking at a column of sediment in side view. Then think about taking a core in search of a tracer whose path is indicated by the figure (see Risk et al., 1978). Also, bear in mind that for the sake of simplicity, we have considered only a one-dimensional random walk. Berg (1993) discusses random walks in two and three dimensions.

As hinted by figure 3.2, particles move at different rates depending on the timescale of observation (see Berg, 1993:10–11). The shorter the

FIGURE 3.2 Pathway of a single particle moving in two dimensions (within the plane of the page) after 18,050 steps. Note that the particle tends to move within certain regions. The distance between points A and B represents 196 step lengths. *Modified from Berg, 1993.*

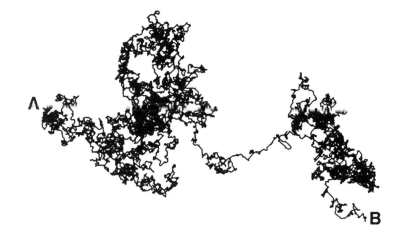

period of observation (t), the greater the particle velocity. The velocity of particles moving within a restricted area (local mixing) is actually much higher than the velocity of particles moving over larger distances (nonlocal mixing) because it takes much longer, on average, to move a long distance than it does a short one, even though the processes involved in moving the particle may be the same. Thus the burrowing activities of animals may occur at *multiple scales* (they are patchy), even in the same species. Johnston (1995) found, for example, that pocket gophers create burrow entrance mounds at spatial scales of decimeters and temporal scales of years, and larger patches (mima mounds) at temporal scales of decades to centuries (approximately circular soil lenses up to about 2 m high and 25–50 m in diameter that occur at densities of 50 to over 100 ha^{-1}). Although certain shallow-water shrimps (such as *Callianassa* and *Upogebia*) may build obvious mounds, such patchiness has not been described for marine organisms.

FLUXES AND FICK

At time t, start with a known number of particles at a fixed point $(x = 0)$ of the x-axis of our example. After the first step (τ seconds), how many (N) of the particles will have moved to the left to point x and how many to the right to point $x + \delta$? What is the net flux of particles across a surface located perpendicular to the x-axis at $x = 0$ after time τ? Because the particles have a 50/50 chance of moving either right or left after each step (τ), for the time τ, half the particles will have moved to the left (to point x) and half to the right (to point $x + \delta$). The net number of particles moving to the right is then

$$-\frac{1}{2}[N(x + \delta) - N(x)].$$ (3.14)

Because a flux is the movement of particles across a surface per unit time, to get the net flux (J) of particles to the right, we divide by the area (A) of the surface and the time (τ)

$$J = -\frac{1}{2}\frac{[N(x + \delta) - N(x)]}{A\tau}.$$ (3.15)

Multiplying by 1 (in the form of δ^2/δ^2) and rearranging terms, equation 3.15 can be rewritten as

$$J = -\left(\frac{\delta^2}{2\tau}\right)\left(\frac{1}{\delta}\right)\left[\frac{N(x + \delta)}{A\delta} - \frac{N(x)}{A\delta}\right]. \qquad (3.16)$$

The term $\delta^2/2\tau$ is the diffusion coefficient (D) and the two terms in the brackets are the concentrations (c) at points $x + \delta$ and x, respectively (the 2 in the denominator is there simply to make the derivation of equation 3.16 a little easier). Equation 3.16 can be simplified to

$$J = -D\left(\frac{1}{\delta}\right)[c(x + \delta) - c(x)]. \qquad (3.17)$$

Because δ is very small, as $\delta \to 0$ (in the limit), equation 3.17 simplifies to

$$J = -D\frac{\partial c}{\partial x}, \qquad (3.18)$$

which is called Fick's first equation after A. Fick, who, in the mid-nineteenth century, modeled diffusion by using the equations of heat conduction earlier developed by French mathematician J. B. Fourier. Equation 3.18 is called a partial derivative (indicated by the use of ∂ rather than d) because the flux actually depends on two independent variables (position x and time t), rather than only one; that is, J is a function of both x and t [$J(x,t)$] rather than of x [$J(x)$] or t [$J(t)$] alone. In the case of a partial derivative, solutions are determined by holding all of the independent variables constant except one (more on solutions involving forms of this equation later).

Fick's first equation is a mathematical statement that the net flux (at x and t) is proportional to the slope of the concentration (at x and t), and is analogous to velocity (the rate of change of position of a particle with respect to time); $-D$ is a constant of proportionality. If particles are uniformly distributed, then the slope $(\partial c/\partial t)$ and J both equal 0. If $J = 0$, then the system is at equilibrium and the particle distribution does not change with time (cf. equation 3.1). If $\partial c/\partial t$ is constant, then J is also constant (see figure 3.3).

Fick's second equation may be derived from the first (refer to figure 3.4). After a period of time τ, $J(x)A\tau$ particles will enter the box (volume $= A \times \delta$) from the left and $J(x + \delta)A\tau$ will leave from the right. If the total number of particles remains constant (particles are neither

FIGURE 3.3 Particles move from right to left because there are more particles on the right than on the left. The flux J (total number of particles that move across distance b per unit time) is equal to the concentration gradient (slope of the line = rise/run = $[C_2 - C_1]/b$) times the diffusion coefficient (D), or Fick's first equation. *Modified from Berg, 1993.*

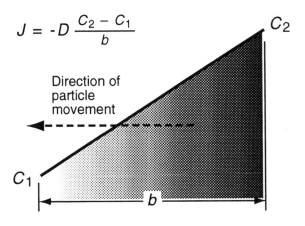

$$J = -D \frac{C_2 - C_1}{b}$$

FIGURE 3.4 Fluxes (J) through a box of thickness δ, used to derive Fick's second equation. Area of the face perpendicular to the movement of particles = A. *Modified from Berg, 1993.*

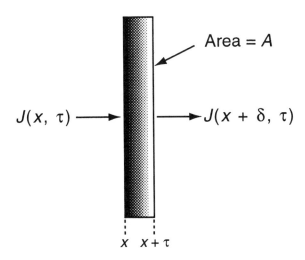

created or destroyed), then the concentration (c) of particles in the box is given by

$$\frac{1}{\tau}[c(t + \tau) - c(t)] = -\frac{1}{\tau}\left[\frac{J(x + \delta)A\tau - J(x)A\tau}{A\delta}\right] \qquad (3.19A)$$

or

$$= -\frac{1}{\tau}\left[\frac{J(x + \delta) - J(x)A\tau}{A\delta}\right], \qquad (3.19B)$$

which simplifies to

$$-\frac{1}{\delta}[J(x + \delta) - J(x)]. \qquad (3.20)$$

As $\tau \to 0$ and $\delta \to 0$,

$$\frac{\partial c}{\partial t} = \frac{-\partial J}{\partial x} = \frac{-\partial}{\partial x}(J). \qquad (3.21)$$

Because $J = -D\partial c/\partial x$ (equation 3.18),

$$\frac{\partial c}{\partial t} = \frac{-\partial}{\partial x}\left(\frac{-D\partial c}{\partial x}\right) = \frac{D\partial^2 c}{\partial x^2}. \qquad (3.22)$$

This is Fick's second equation. It states that the rate of change of concentration (at x and t) is equal to the rate of change of the slope (the curvature) of the concentration function at x and t (in other words, how much the slope is changing at each instant); D is a proportionality constant. This equation is analogous to that of acceleration (the rate of change of velocity with respect to time, or the second derivative of the change in particle position with respect to time). If the slope of the curve at some instant is constant, then $\partial^2 c/\partial x = 0$, $J = 0$, and the system is at equilibrium (the particle distribution does not change): Just as many particles diffuse from the region of higher concentration into the system (box) as out of the box to the region of lower concentration. Fick's second equation can be used to determine how an initially nonuniform distribution of particles will redistribute itself through time. We have now derived the first term of the Guinasso–Schink equation.

Derivation of the second term of equation 3.1 is much shorter, at least for our purposes. If all the particles in a particular distribution move (drift) in a positive direction with (burial) velocity v, then the flux of particles at point x must increase by $vc(x)$ and Fick's first equation becomes

$$J = \frac{-D\partial c}{\partial x} + vc. \qquad (3.23)$$

This equation is used to rederive Fick's second equation:

$$\frac{\partial c}{\partial t} = \frac{D\partial^2 c}{\partial x^2} - \frac{v\partial c}{\partial x}. \qquad (3.24)$$

By adding the term λc for radiotracer decay, we arrive at the Guinasso–Schink equation (3.1).

SOLUTION OF THE GUINASSO–SCHINK EQUATION

The solution to equation 3.1 is

$$C = C_0 \exp\left[\frac{v - (v^2 + 4\lambda D)^{1/2}}{2D} x\right], \qquad (3.25)$$

where C_0 is the original tracer concentration (Berner, 1971:102). There are at least two ways to explain how equation 3.25 is derived: One involves the determinants and eigenvalues of a set (matrix) of coefficients for two equations solved simultaneously (see Causton, 1987: 269–276, or Beltrami, 1993, appendix B for a brief treatment; you may have to consult a statistics text, such as Davis, 1986, chapter 3, as I did several times, to review the concepts of determinant and eigenvalue).

The other approach is simpler and will be used here. Assume that $y = e^{\psi x}$, which is a common general solution of a linear first-order differential equation, is also a solution to equation 3.1, which is a linear second-order differential equation. Then $y' = \psi e^{\psi x}$ (first derivative or dy/dx) and $y'' = \psi^2 e^{\psi x}$ (second derivative or $d^2 y/dx^2$), because the derivative of an equation involving the base of natural logarithms (e) takes the form $dy/dx = du \times e^u$, where $u =$ the entire exponent. For example, the first derivative (dy/dx) of $y = e^{2x}$ is $2e^{2x}$ and the second derivative is $4e^{2x}$ or 2^2e^{2x}). Substituting into equation 3.1,

$$D(\psi^2 \exp^{\psi x}) - v(\psi \exp^{\psi x}) - \lambda(\exp^{\psi x}) = 0. \qquad (3.26)$$

Factoring out $e^{\psi x}$,

$$\exp^{\psi x}[D(\psi^2) - v(\psi) - \lambda] = 0. \qquad (3.27)$$

Let $D = a$, $-v = b$, and $-\lambda = c$. Then, using the general solution for a quadratic equation,

$$x = \frac{-b \pm \sqrt{b^2 - 4ac}}{2a}, \qquad (3.28)$$

we can solve for ψ by substitution:

$$\psi_1, \psi_2 = \frac{-v \pm \sqrt{v^2 - 4\lambda D}}{2D}. \qquad (3.29)$$

The two values of ψ are the eigenvalues of the matrix of equations. One of these eigenvalues results in a solution that contains the square root of a negative number (also called an imaginary number). Fortunately, we use symmetric matrices, in which the number of columns equals the number of rows and the solutions to which are real eigenvalues (cf. equation 3.25).

ESTIMATION OF SEDIMENTARY PARAMETERS

In order to solve the Guinasso–Schink equation (3.1), we have to estimate values for the sedimentary parameters (that is, we must determine what are called initial or boundary conditions). Of course, the radioactive decay coefficient depends on the choice of radiotracer. Which radiotracer should we choose? This depends on how deep one is working in the shallow sediment column beneath the sediment–water interface or SWI (and the amount of funding available). For example, core X rays and ^{210}Pb (half-life $\approx$ 22.3 yr) have been used extensively to estimate the thickness of the surface mixed layer (m) or taphonomically active zone (chapter 4, figure 4.5) from the change in slope of the concentration profile (Robbins, 1978; Benninger et al., 1979; Officer,

1982). ^{210}Pb is produced mainly by natural decay of the gas ^{222}Ra, which diffuses out of the earth's crust into the atmosphere, and is rapidly scavenged from the air by particulates settling through the atmosphere. Excess ^{210}Pb levels are supported in the mixed layer by continued input from the atmosphere. Typically, the upper portion of the ^{210}Pb profile is fairly constant within the mixed layer, but below the mixed layer ^{210}Pb is no longer supported and ^{210}Pb decays to lower levels with depth.

The apparent sedimentation rate (v') can be determined for the last 100 years from the slope of the natural logarithm (ln) of ^{210}Pb concentration versus depth (x) below the mixed layer. v' may confound true sedimentation and mixing by bioturbation, however, and may therefore overestimate the true value of v, depending on the magnitude of bioturbation, as estimated by D (Officer, 1982). Instead, longer-term or "true" v can be estimated from the downcore profile of ^{14}C (half-life ≈ 5,730 yr) for the past few thousand years. This is also an apparent v, but the correction is negligible for high v rate areas such as coastal regions (Officer, 1982).

Even shorter-term sedimentation rates can be determined using ^{137}Cs. Unlike ^{210}Pb and ^{14}C, ^{137}Cs (half-life ≈ 30 yr) is a time-dependent (impulse) tracer derived from atomic weapons testing, which peaked around 1963 or 1964 (Robbins, 1982), and has been used to estimate v for the last 20 years. Although it may be mobile because of changes in pore water redox chemistry caused by organic matter decay (T. M. Church, personal communication), Sharma et al. (1987) successfully modeled ^{137}Cs profiles in studies of marsh sediments.

^{7}Be is not as susceptible to geochemical changes (T. M. Church, personal communication; see also Lord and Church, 1983; Luther et al., 1986, 1991), and can be used to estimate very short-term sedimentation rates (on the order of a few months; Krishnaswami et al., 1980; Sharma et al., 1987). ^{7}Be is formed by spallation reactions of cosmic ray particles with atmospheric nitrogen and oxygen (Lal and Peters, 1967) and has a half-life of about 28 days (T. M. Church, personal communication).

The choice of radiotracer at the outset is also critical because the sedimentary parameters are all interrelated; the determination of one affects the value of another (see also Officer, 1982; Boudreau, 1986a, 1986b). For example, Aller and Cochran (1976) estimated D in shallow subtidal sediments of Long Island Sound. They used the short-lived radiotracer ^{234}Th (half-life ≈ 24 days), which is a decay product of ^{238}U and is most concentrated in the uppermost 5 cm of the sediment column beneath the SWI (it decays too rapidly to normally penetrate much deeper), to estimate the biodiffusion (reworking) rate (D) from the equation $D = mv'$ (where m = mixed layer thickness in centimeters; more on the derivation and limitations of this equation later). They assumed that because ^{234}Th has such a short half-life, the apparent rate of sediment accumulation v' times m is approximately D (that is, $D >> v$ for short intervals of time). Later, Benninger et al. (1979) used these ^{234}Th-based calculations for D, along with D values based on ^{239,240}Pu for depths up to 10 cm, and a v' based on ^{210}Pb. They then used a "composite-layer, mixing + sedimentation model," in which D varied in discrete layers with depth, to determine which model best fit the data (cf. figure 4.5). Using a similar set of composite-layer models, Officer (1982) calculated the mixing parameter D from

$$D = \frac{(v' - v)v'}{\lambda};$$
(3.30)

he obtained estimates of D similar to those published by other workers for deep sea cores and the shallow subtidal of Long Island Sound using ^{210}Pb.

In the case of conservative (nondecaying), instantaneous ("impulse") tracers (those that produce "event" layers and do not decay, such as volcanic ash or microtektites), the term λ of equation 3.1 may be ignored. In fact, microtektite data were first used by Guinasso–Schink (1975) to test the diffusion model of sediment mixing. Officer and Lynch (1983) derived a new set of solutions to equation 3.1 because they found that the Guinasso–Schink model was in error. Solutions to their equations are rather formidable and are found using a technique called optimization, in which an iterative technique called the Newton–Raphson method is used to find successively better approximations to the solutions. This technique is not discussed further here (see Cutler, 1993, for a concise review); instead, a simpler approach by Officer and Lynch (1983) is discussed.

By making the variables in equation 3.1 dimensionless (unitless), Guinasso and Schink (1975) concluded that mixing may be described by a dimensionless parameter G, where

$$G = \frac{D}{mv}.$$
(3.31)

When G is small (≤ 0.1, as in rapid sedimentation or slow mixing), an impulse tracer is buried below the mixed layer before it can be extensively reworked (that is, its concentration distribution is bell-shaped or Gaussian; see figure 3.5 and compare to figure 3.1 turned on its side). When G is large (≥ 10; slow sedimentation or rapid mixing), the tracer exhibits an exponential decrease in concentration with time in the mixed layer, and a uniform distribution in the mixed layer (figure 3.5). For values of $G \geq 0.6$, Officer and Lynch (1983) developed a simpler method for calculating D, m, and the original depth of deposition of the tracer using

$$G = \beta_2 + \frac{\sqrt{\beta_2^2 + \pi^2 \beta_1 (\beta_2 - \beta_1)}}{2\pi^2 \beta_1} \qquad (3.32)$$

and

$$m = \frac{1 + 1/4G}{\beta_1}, \qquad (3.33)$$

where β_1 and β_2 are the slopes (fitted by linear regression) of the conservative tracer concentration profile in sediment. D is then calculated from equation 3.31 using v obtained from radiotracers. The depth of the tracer (impulse) layer (or other stratigraphic signal) that would have been observed in the core without bioturbation (x_0) can then be found from the observed (or interpolated) tracer peak x_m by

$$x_0 = x_m - \left(1 - \frac{\ln \left[(8\pi^2 G^2 + 2)/(4G + 1)\right]}{\pi^2 G - 1}\right) m \qquad (3.34)$$

CAVEATS OF DIFFUSION-BASED MIXING MODELS

Wheatcroft (1990) criticized the parameter G on a number of grounds. First, the G values for most environments lie between 0.1 and 10, not at either extreme (see also Officer and Lynch, 1983). Second, equation 3.31 is based on what is called a Péclet number (actually, G is the inverse of a Péclet number), which is used in engineering to compare the relative importance of advection versus diffusion in the transfer of heat or mass away from an object. Although G, like a Péclet number,

FIGURE 3.5 Relationship between mixing parameter *G* and conservative impulse tracer concentration profiles (such as microtektites or volcanic ash) at different times. Depth in mixed layer units; time normalized to time required for one mixed layer thickness to accumulate. When *G* is small (rapid sedimentation, slow mixing), an impulse tracer is buried below the mixed layer before it can be extensively reworked (bell-shaped profile). If *G* is large (slow sedimentation, rapid reworking), the concentration of tracer decreases upward exponentially, but its concentration in the mixed layer in this case is fairly uniform. Compare with figure 3.1 turned on its side. *From R. E. Martin, 1993. Time and taphonomy: actualistic evidence for time-averaging of benthic foraminiferal assemblages. In S. M. Kidwell and A. K. Behrensmeyer, eds.,* Taphonomic Approaches to Time Resolution in Fossil Assemblages, *pp. 34–56. Paleontogical Society Short Courses in Paleontology No.6. Reprinted with permission of the Paleontological Society.*

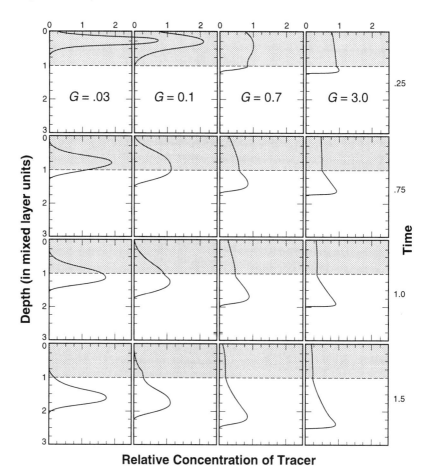

consists of a diffusion coefficient, a (burial) velocity, and a length, Wheatcroft (1990) feels that G is not dynamically correct. For example, if m increases, then for a given burial velocity, it takes longer for a tracer signal to move through the mixed layer; therefore, the tracer is more likely to be dispersed and the signal destroyed. But according to equation 3.31, if m increases, G should decrease and the tracer signal should be better preserved (figure 3.5). Wheatcroft (1990) suggested another approach (also based on diffusion) to measuring the degree of dissipation (spreading) of an impulse tracer (cf. figures 3.1 and 3.5):

$$c = \frac{1}{2}\left[\mathrm{erf}\left(S/2\sqrt{DT_m}\right)\right], \qquad (3.35)$$

where S = signal layer thickness, T_m = transit time or the time for the signal to traverse through the mixed layer, and erf is the error function, which depends on the function involved and is tabulated in various engineering texts (such as appendix II of Carslaw and Jaeger, 1959; see chapter 2 of Crank, 1975, and Berg, 1993, for more on equation 3.35). According to Wheatcroft (1990), equation 3.35 yields more realistic values for dissipation of an impulse signal.

Another criticism of diffusion-based models is that bioturbation is not strictly diffusive. Remember that D is defined as $\delta^2/2\tau$. D depends on the square of the distance that particles are moved (step length = δ) and inversely on the time (τ) between movements. What this means for sediment particles (as opposed to molecules in a fluid; refer to the previous discussion of movement of the lysozyme molecule) is that they stay in one place for long periods of time before being moved to another spot, where they remain for some time, and so on (Wheatcroft et al., 1990; see figure 3.2). Large deposit feeders are more likely to dominate mixing over greater distances than diffusive processes that act over much shorter scales (cf. figure 3.2). Biodiffusion may be dominated, for example, by advective mixing or apparent burial (as opposed to sediment accumulation from the water column at the SWI) caused by bioturbating organisms such as conveyor-belt deposit feeders (CDFs; various species of worms and shrimp). CDFs move sediment much more rapidly than diffusion alone, as these organisms are constantly transferring sediment downward (apparent burial) and then upward to the SWI in conveyor-belt fashion (van Straaten 1952; Rhoads and Stanley 1965; Meldahl, 1987). Bioturbators may also transfer whole parcels of sediment (not just individual sediment grains) intact into burrows in or below the mixed layer (nonlocal mixing). Moreover, rapid horizontal mixing coupled with vertical advection can produce tracer profiles characteristic of diffusion models by transporting sediment grains away from zones of vertical transport. Nevertheless, the net result of these activities is that the tracer profile may still appear to have resulted from diffusion alone, no matter what the actual process of sediment movement is, and can seem to be described by a diffusion coefficient (D; Boudreau, 1986a, 1986b; Boudreau and Imboden, 1987; Wheatcroft et al., 1990).

Despite the limitations of the G model, Hippensteel and Martin (1998) calculated mean G values of 0.032 $\pm$ 0.046 and 0.259 $\pm$ 0.384 for the high and low marsh behind barrier islands off Charleston, S.C. They estimated β_1 and β_2 of South Carolina overwash fans by measuring the slopes of obviously bioturbated (smeared upward) peaks of Oligo-Miocene foraminifera that had been deposited by storms in the marshes behind barrier islands. Despite the wide variation in G values, their G values agree with the trend of increasing bioturbation from high to low marsh calculated by Sharma et al. (1987) based on equation 3.1.

SIZE-SELECTIVE FEEDING

Not all particles are created equal, especially when it comes to size. Many deposit-feeding (bioturbating) and suspension-feeding organisms have an intricate sorting apparatus and behavior patterns that are used to discriminate between sediment particles. Particles that are too large are often rejected before they are ingested because their size indicates that most likely they will provide little or no nutriment.

Surprisingly, the phenomenon of size-selective feeding has been rejected out of hand by many workers as an important influence on tracer (fossil) distributions and temporal resolution at small scales. Based on studies of volcanic ash and microtektite distributions, Ruddiman et al. (1980) concluded that particle size selectivity by bioturbators is unimportant in the deep sea (but see dissenting view of Glass in Ruddiman et al., 1980; curiously, Ruddiman and Glover, 1972, had earlier concluded that bioturbation was important in controlling ash distributions in deep-sea cores). Wheatcroft and Jumars (1987; see also Wheatcroft, 1992) statistically reanalyzed the data of Ruddiman et al. (1980), however, and concluded that given a particle size range of 11–

500 μm (which includes the range of volcanic ash), small particles are more likely to be mixed than large ones, and that larger microfossils (such as foraminifera) are less likely to be advected than smaller ones (coccolithophorids).

Robbins (1986) also concluded (based on models of freshwater ecosystems with the oligochaete worm *Tubifex tubifex*) that particle-selective feeding can produce nonuniform tracer profiles. Rejection of a tracer by these organisms (which are CDFs) reduces the surface concentration of the tracer and produces biogenic graded bedding (see also chapter 4), whereas selective ingestion and transport of the tracer to the SWI increases the tracer concentration there. Thus increasing sedimentation rate and the selectivity factor (that is, discrimination against large particles) decreases the displacement and homogenization of large particles, whereas increasing feeding rate (on small particles) and depth of feeding tends to increase downward displacement and homogenization of small tracer particles (Robbins, 1986). Repeated cycles of ingestion and redeposition of a tracer at the SWI result in subsidiary tracer peaks below the SWI. Robbins's (1986, his figure 3) model-generated diagrams of tracer distributions generated by particle-selective feeding bear a strong resemblance to the distribution of ash layers published by Ruddiman and Glover (1972, their figure 8).

Unfortunately, diffusion-based models are about all that is available at this point (Wheatcroft's [1990] approach excepted). Fortunately, though, they also provide a beginning for resolving processes recorded in the fossil record, as discussed in chapter 4. Numerical modeling has also begun (Pizzuto and Schwendt, 1997).

One final note before moving on: The intensity of bioturbation has not remained constant through the Phanerozoic. It has apparently increased in rate and moved progressively from nearshore to offshore environments. This has important implications for the cycling of geologically and biologically important elements such as carbon and phosphorus, as discussed in chapter 7.

4

Time and Taphonomy

Ay, but to die, and go we know not where;
to lie in cold obstruction and to rot;
This sensible warm motion to become
A kneaded clod.
—*William Shakespeare,* Measure for Measure

Death has shaken out the sands of thy glass.
—*John Brainard,* Lament for Long Tom

Taphonomy, which literally means the "laws of burial," is the science of the formation of fossil assemblages; it is concerned with the processes of preservation and how they affect information in the fossil record (Behrensmeyer and Kidwell, 1985). Not all fossils are created equal, and their differential preservation as they are incorporated into a fossil assemblage plays an extremely important, but still largely unappreciated role (despite the burgeoning literature) in determining the temporal resolution of fossil assemblages.

TIME AVERAGING AND TEMPORAL RESOLUTION

One of the primary artifacts overprinted on the fossil record is the phenomenon of time averaging. This process is responsible for mixing fossils of different ages into a single assemblage and results from several factors. Time averaging is a result of both bioturbation (chapter 3) and physical reworking by waves and currents (as in storms). Even without these processes, though, time averaging would still occur because there is an inherent bias in the fossil record: The generation times of organisms are typically much shorter than net rates of sediment accumulation and burial (Kidwell, 1993a; figures 4.1, 4.2, and

4.3). Because of time averaging, a fossil assemblage may represent a minimal duration of a few decades at best, and more likely hundreds to thousands of years (figure 4.4), unless the assemblage is rapidly preserved by unusual conditions (such as *Lagerstätten*). Perhaps the most famous example of a Lagerstätte is the Burgess Shale (Cambrian) of British Columbia, in which soft-bodied organisms, not normally preserved because they lacked hardparts, were buried suddenly by a submarine landslide and preserved as carbon films.

Time averaging is a relative term, however, and depends on the scale of the phenomenon of interest (Kowalewski, 1996). Short-term population (high-frequency) phenomena are often lost (see also chapter 3), but that is advantageous if one is interested in longer-term processes. For example, some death assemblages appear to be comparable to the results of repeated biological surveys that document changes in community species composition and diversity over several decades or

FIGURE 4.1 Formation of fossil concentrations by changes in net sediment accumulation. Shell input is held constant in this model. *Modified from Kidwell, 1986.*

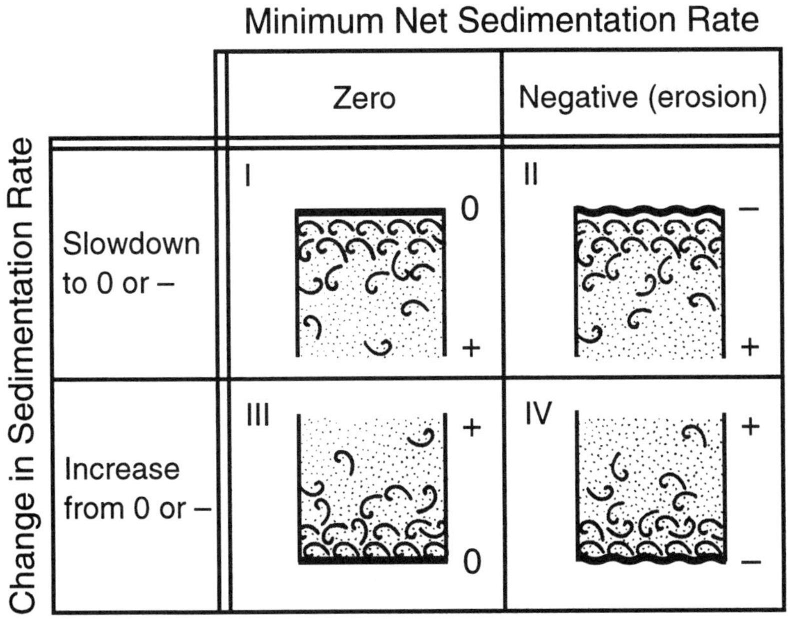

FIGURE 4.2 Predicted taphonomic effects resulting from changes in sediment accumulation and time averaging. Compare with figure 4.1. *Modified from Kidwell, 1986.*

R-sediment model: Predicted postmortem bias

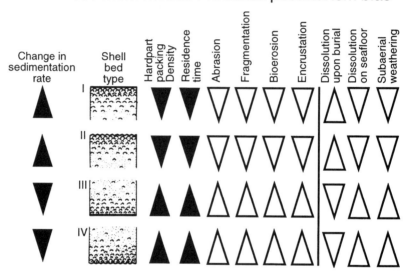

FIGURE 4.3 Exposure versus transportation effects. *From S. M. Kidwell, 1993a. Time-averaging and temporal resolution in Recent marine shelly faunas. In S. M. Kidwell and A. K. Behrensmeyer, eds.,* Taphonomic Approaches to Time Resolution in Fossil Assemblages. *Paleontological Society Short Courses in Paleontology No. 6:9–33. After Johnson, 1960. Reprinted with permission of the Paleontological Society.*

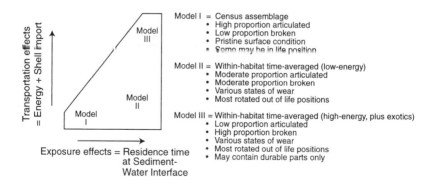

FIGURE 4.4 Estimated durations of time averaging for various types of fossil assemblages. *Modified from Kidwell and Behrensmeyer, 1993.*

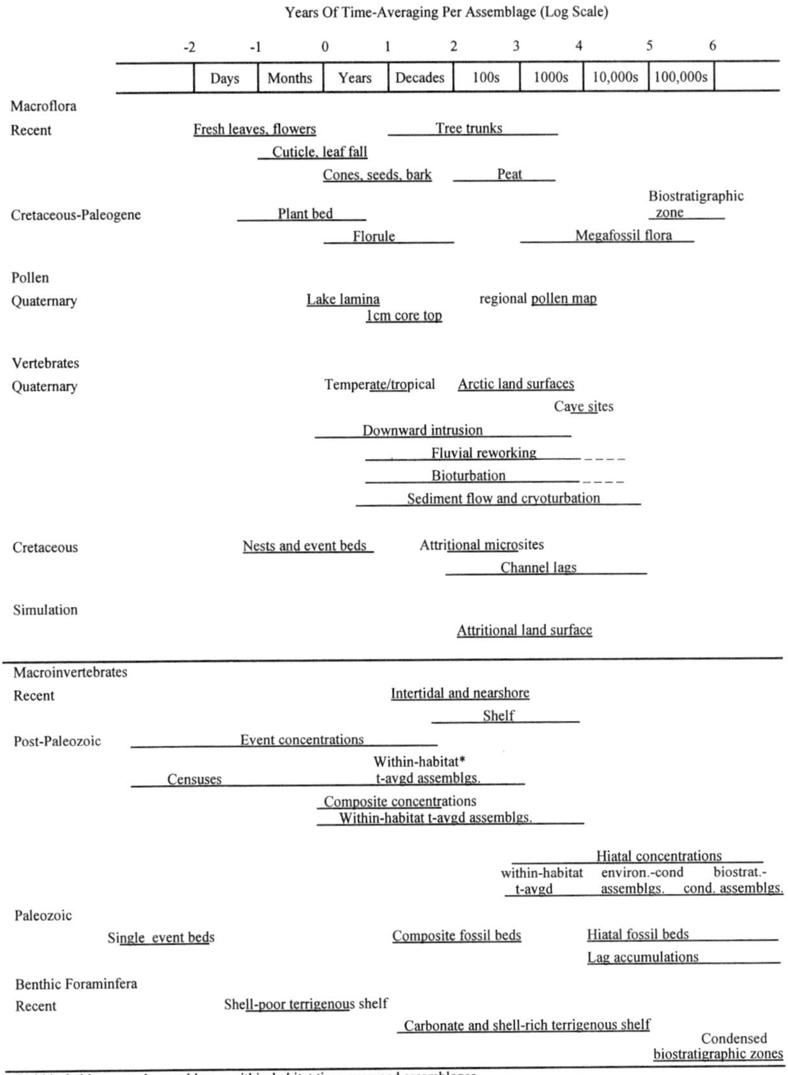

*within-habitat t-avgd assemblgs. = within-habitat time-averaged assemblages
environ.-cond assemblgs = environmentally - condensed assemblages
biostrat. cond. assemblgs = biostratigraphically condensed assemblages

more, including sudden phenomena that might be missed by short-term sampling regimes (Kidwell and Bosence, 1991; see also chapter 5). Although time averaging becomes less significant over longer scales of measurement, the incompleteness of the fossil record looms ever larger (Sadler, 1981; Sadler and Strauss, 1990; Kowalewski, 1996; see also chapter 2).

Between these two extremes there is a threshold of time averaging beyond which certain phenomena are lost. Thus a fossil assemblage that has been time-averaged on one scale may not be time-averaged on another. For example, if the duration of time averaging, as expressed by the difference between the ages of the oldest and youngest shells in an assemblage, is about 1000 years and the phenomenon occurs over a scale of 10,000 years, then time averaging is insignificant. On the other hand, if the desired scale is 100 years, the assemblage is probably a poor choice for study (Kowalewski, 1996; see also chapters 1 and 2).

In order to further examine the effects of time averaging on temporal resolution, consider yet another sediment mixing model. In the case of slow sedimentation or rapid mixing (see the discussion of G in chapter 3), the Guinasso–Schink diffusion model is identical to the sediment mixing model developed by Berger and Heath (1968). This "box" model assumes that the shallow sediment mixed layer is reworked so rapidly with respect to sedimentation that the concentration of a tracer in the mixed layer is uniform before it is finally entombed in the historical layer underneath (figure 4.5). On the surface, this seems to be a reasonable assumption for deep-sea cores (for which this model was developed) because the mixed layer is usually thin (5–10 cm) and rates of deep-sea sediment accumulation are slow (a few cm/thousand years [ka] on average). This means that, in most cases, organisms ought to have sufficient time to homogenize the mixed layer in the deep sea, but this is not always the case. Indeed, just because sediment layers in a core do not look like they have been bioturbated does not mean that they have not been disturbed (Berger et al., 1979).

I start with a conservative impulse tracer, such as a microtektite or volcanic ash layer, both of which are event layers. According to the Berger–Heath model,

$$\frac{dP}{P} = \frac{-ds}{m},$$

(4.1)

FIGURE 4.5 Hypothetical plots of mean age of a tracer versus depth in sediment. (a) No mixing. Slope of line is constant and determined by sediment accumulation rate only. (b) Box model (complete mixing within mixed layer; D is effectively infinite). Mean tracer age is the same at all depths within mixed layer. Age profile of tracer within historical layer is displaced toward younger mean age relative to unmixed profile (*dotted line*). (c) Diffusive mixing within mixed layer. Slope of age profile within mixed layer is proportional to D. Age profile is displaced, although not as much as in (a). (d) Mixed layer consists of two layers with different values of D. Mixing is complete in the upper layer, but D has an intermediate value in the lower mixed layer. *Modified from Cutler, 1993.*

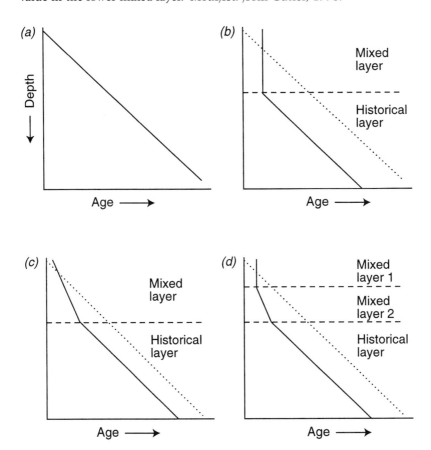

where
P = probability of finding a particle in the mixed layer,
s = thickness of sediment deposited on top of the mixed layer, and
m = mixed layer thickness.

Integrating equation 4.1 results in a decay formula,

$$P = P_o \exp^{(-s/m)} . \tag{4.2}$$

Thus, after deposition of the event layer, the stratigraphic distribution of the tracer is given by

$$P = P_{oz} \exp^{(s_e+m)/m} , \tag{4.3}$$

where
P_{oz} = original concentration of z at a distance of m *below* the top of the layer, and
s_e = thickness of sediment deposited after the event layer has been deposited.

In other words, the concentration of the tracer above its true level of deposition decreases exponentially according to the thickness of sediment deposited after the layer has been deposited.

Based on equation 4.3, Berger and Heath (1968) calculated the proportion of the original concentration of a tracer found in the sediment at a distance s_e above the true layer. By specifying acceptable levels of reworking (contamination), the level of stratigraphic resolution may be calculated. In their example (figure 4.6), if m = 4 cm and the "contaminant" level (due to reworking) is 10 percent of the original concentration of the tracer, then the original depth of deposition may be as much as 9 cm deeper in the section. For a sedimentation rate of 1 cm/ka, this means a resolution of 9000 yr; for a rate of 4 cm/ka, a resolution of 2250 yr is obtained. The exact level of deposition is indicated when the tracer concentration is $1/e$, or 0.37 times its maximum concentration, and is independent of m (Berger and Heath, 1968). This is because at this level, s_e = 0 and the exponent in equation 4.3 becomes -1. Therefore, the right-hand side of equation 4.3 becomes $P_o e^{-1}$ or $P_o(1/e)$. (The value $1/e$ is also typically used as a measure of the return time of a system to its original state after a disturbance). A similar model was developed by these authors for species appearances (smeared downward). Berger and Heath (1968) concluded that serious stratigraphic errors may occur if the stratigraphic range of a species is similar in thickness to the mixed layer (figure 4.7); this is most likely to occur in the deep-sea during times of climatic fluctuations, and when species may alternately appear and disappear.

This same phenomenon was observed earlier by Glass (1969). He found that in deep-sea cores (mainly Indo-Pacific), microtektites were dispersed over intervals of 35–90 cm (average 60 cm), representing 33–320 (average 120) ka. Stratigraphic distributions seemed to be con-

FIGURE 4.6 Proportion of the original concentration of a tracer (P/P_o) found at a distance (s_e) in sediment above its level of disappearance. In this example, if the thickness of the mixed layer (m) = 4 cm and upward mixing has taken place to a distance 9 cm above the original impulse layer, then the concentration of the tracer has decreased to 0.1 of its original value. *Modified from Berger and Heath, 1968.*

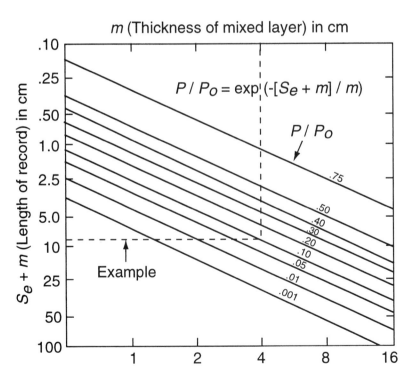

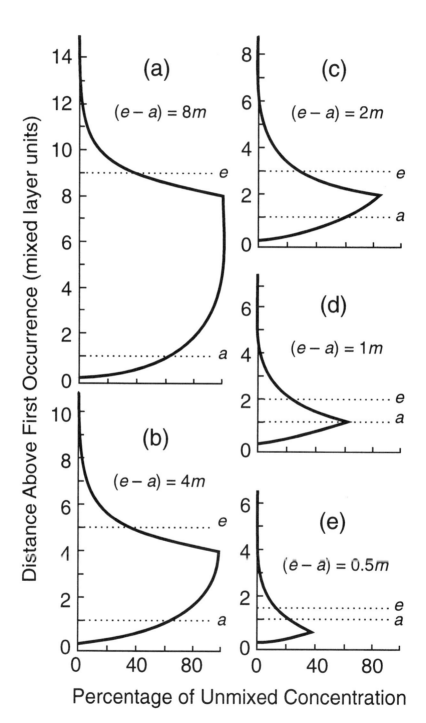

trolled by the intensity of burrowing (estimated by burrow frequency), and larger microtektites were concentrated in the upper half of the bioturbated impulse layer in some cores (see also chapter 3). Extrapolating to fossils, he concluded that the apparent time of extinction of two fossil species of different average size that became extinct simultaneously might be significantly altered by bioturbation. Based on abundance curves, he also concluded that the first appearance of a species is a more reliable stratigraphic marker than its extinction (remember that Oppel also used first appearances; see chapters 2 and 3). Recently, MacLeod and Huber (1996) concluded that at the Cretaceous/Tertiary (K/T) boundary, up to 30 percent of the mass of foraminifers in a sample may derive from reworked specimens (see also chapters 2 and 8).

TAPHONOMIC HALF-LIVES AND TAPHOFACIES

So far, we have discussed the concentration and distribution of a conservative tracer that does not decay with time. What about fossils, which are used to determine temporal resolution but can be destroyed by abrasion or dissolution? In these cases, a term analogous to λ of equation 3.1 must be determined. Unlike radiotracers, however, the half-life of a calcareous microfossil depends on the environment in which it is deposited (Martin, 1993). Shell λ, in turn, affects the abundance, preservation, probability of reworking, and temporal resolution of foraminiferal assemblages, and depends on such factors as sediment accumulation rate (preservation is enhanced by rapid burial); macro- and microbenthic shell inputs and total $CaCO_3$ content of sediment (which buffers against dissolution); alkalinity (a measure of the total dissolved carbon in seawater, where $\Sigma CO_2 = [HCO_3^-] + [CO_3^{2-}]$), which also buffers against dissolution when sufficiently high; shell size (surface/volume ratio), mineralogy, microstructure and organic matter content, all of which may affect chemical reactivity (see Corliss and Honjo, 1981; Bremer and Lohmann, 1982; Kotler et al., 1992;

FIGURE 4.7 Tracer distributions resulting from bioturbation in relation to their appearance (a), extinction (e), and thickness of the mixed layer (m). Distance above appearance on the x-axis is given in terms of mixed layer units. As the stratigraphic range of a marker decreases through the sequence of diagrams, stratigraphic errors become more serious. *Modified from Berger and Heath, 1968.*

and Green et al., 1993a); antecedent topography, which may influence sediment thickness and the depth of bioturbation; and habitat or depth at which the organism lives, which depends on such factors as organic matter and oxygen content of the sediment.

Fossil λ also depends on the rate of bioturbation. Bioturbation causes the formation of carbonic and sulfuric acids through the oxidation of organic matter and sulfides, respectively (Walter and Burton, 1990):

$$CH_2O + O_2 \rightarrow H_2CO_3, \tag{4.4}$$

$$2O_2 + HS^- \rightarrow SO_4^{2-} + H^+ \tag{4.5}$$

Conversely, bacterial SO_4^- reduction enhances pore water alkalinity and therefore $CaCO_3$ preservation (Walter and Burton, 1990):

$$2CH_2O + SO_4^{2-} \rightarrow HS^- + HCO_3^- + H_2CO_3 \tag{4.6}$$

(see also Canfield and Raiswell, 1991, for review).

Besides affecting the preservation of dead hardparts in the mixed layer, pore water chemistry affects the distribution of animals living below the SWI (infauna) and the preservation of their dead shells. Benthic foraminifera, for example, exhibit microhabitat preferences that reflect surface productivity, organic carbon influx (from surface photosynthesis either at the SWI or from the water column above), and pore water oxygen content. Corliss and Emerson (1990) concluded that habitat depth is related to organic carbon input, which decreases with depth, and dissolved oxygen in the pore waters (based on dissolved Mn^{2+} profiles; pore water concentrations of Mn^{2+} decrease with increasing O_2 because Mn^{2+} precipitates in the presence of O_2). Shelf species tend to be epifaunal (shallow oxic layer), whereas on the continental slope and rise, the oxidized surface layer is thick and foraminifera are primarily deep infaunal species. At greater depths, deep-water epifaunal species predominate because organic carbon (food) becomes limiting (see also Corliss, 1985, 1991). Loubere (1991) found that certain deep-sea (East Pacific Rise) foraminifera respond to productivity gradients. Significantly, some species appear to be highly adaptable (in terms of microhabitat preference and depth stratification) to changes in food availability and environmental conditions (Linke and Lutze, 1993). McCorkle et al. (1990) found that in deep-sea benthic foraminifera, carbon isotope composition, which is commonly used to infer past changes in oceanic productivity (see chapters 7 and 8 of this book), can be influenced by microhabitat preferences and pore water chemistry.

Loubere (1989; see also Loubere et al., 1993a, 1993b, and 1995) also found that significant variation in species abundances occurs in the mixed layer as a result of taxon depth stratification. Infaunal taxa increase nonlinearly in abundance with depth to the level of their true habitat, below which taxon abundance remains constant, whereas epifaunal assemblages may be significantly modified. He concluded that the best representation of total taxon abundance occurs at the base of the mixed layer. Denne and Sen Gupta (1989) came to similar conclusions; that is, surface grab samples, which have often been the basis of foraminiferal distribution studies, are not necessarily representative of subfossil assemblages of open marine environments. Goldstein and Harben (1993) and Goldstein et al. (1995) concluded that deep (up to 30-cm) infaunal populations of salt marsh foraminifera can affect paleoenvironmental interpretations and therefore the construction of Holocene sea-level curves.

Differences in preservational gradients, extending from either the SWI into the sediment or across continental margins (shelf and slope), also strongly influence time averaging and stratigraphic resolution. Each depositional environment may be characterized by its own taphonomic (preservational) setting or taphofacies (see Speyer and Brett, 1988) that depends on rates of sediment accumulation, bioturbation, sulfate reduction, and so on.

Two case studies for shallow nearshore settings (shallow subtidal carbonates at Discovery Bay, Jamaica, and siliciclastic tidal flat sediments of the northern Gulf of California) are presented in this chapter to demonstrate the complex interplay of taphonomic factors in preserving and time-averaging fossil assemblages in shallow-water environments. These investigations were chosen because they are among the best-documented marine examples and because the results demonstrate that even some of our simplest assumptions regarding the behavior of fossil particles are erroneous. Deeper shelf and slope settings are also discussed.

DEATH, DECAY, AND TEMPORAL RESOLUTION IN NEARSHORE SETTINGS

Dead tests of large reef-dwelling foraminifera, such as *Archaias* and *Amphistegina*, found in sediment assemblages often exhibit considerable surface degradation. Based on experimental analyses of both abrasion and dissolution resistance of these and other modern reef-dwelling foraminifera from Discovery Bay, Jamaica, Kotler et al. (1992) concluded that, once produced, foraminiferal tests in pure carbonate environments tend to persist for long periods of time, although at that time they were uncertain exactly how long (years? decades? centuries?).

Later, accelerator mass spectrometer (AMS) ^{14}C dating of mildly to highly corroded tests of *Archaias* and *Amphistegina* produced results of 474–2026 years and 936–1839 years, respectively! Moreover, pristine tests ranged from modern (as indicated by "post-bomb" ages) up to 2000 years old, thereby exceeding the ages of mildly or highly corroded tests (Martin et al., 1996b). High levels of $CaCO_3$ in pure carbonate environments appear to buffer against dissolution, and thus promote extensive mixing of old tests and much younger shells while inhibiting the formation of taphonomic damage on the tests themselves.

Kotler et al.'s (1992) results were in marked contrast to those of Walter and Burton (1990), who found that red algal (18 mol% Mg calcite), echinoid (12 mol% Mg), and coral (aragonite) substrates dissolved rapidly in field experiments. The carbonate substrates used by Walter and Burton (1990) would appear to be far more reactive, however, than foraminifera (Kotler et al., 1992). Thus differential dissolution of carbonate particles, *even in carbonate environments*, results from differences in mineralogy. Other factors involved no doubt include substrate size, shape, and surface area (surface/volume ratio, high in red algae and low in foraminifera); organic matter content (also high in algae); and microstructure.

Preservation in shallow-water siliciclastic regimes, such as those of Choya Bay in the northern Gulf of California, has also been surprising. Flessa et al. (1993; see also Flessa and Kowalewski, 1994) found that disarticulated bivalve shells (*Chione*) that had been collected from the SWI of Choya Bay exhibited a broad range of surface degradation, and that shell condition (taphonomic grade) was not an infallible indicator of shell age (time since death); moderately old specimens (about 1800 years) were sometimes well preserved, whereas younger shells (several hundred years) were sometimes more highly degraded (Flessa et al., 1993). Flessa et al. (1993) suggested that a shell's surface condition indicates primarily the residence time of the shell at the SWI, not its age.

Martin et al. (1996b) arrived at similar findings, with one important exception. Amino acid racemization analyses (alloisoleucine/isoleu-

84

cine or A/I ratios, which are used to date shells beyond the range of ^{14}C dating) of *Chione* indicated that the scale of time averaging of bivalve assemblages at Choya Bay may be one to two orders of magnitude greater (~80,000–125,000 years) than that reported by Flessa et al. (1993). Of 17 *Chione* valves analyzed by Martin et al. (1996b), 9 fell into this age range; the others were at most a few thousand years old (based on AMS dates). Significantly, 6 of these 9 valves were indistinguishable in appearance from much younger *Chione* valves. This scale of time averaging is not unique to Choya Bay. Wehmiller et al. (1995) found extensive age mixing of Pleistocene and modern bivalves (mainly *Mercenaria*) at 21 beach sites between New Jersey and Florida.

Based primarily on size differences between foraminifera and *adult* molluscs, Martin (1993) predicted that larger (and presumably more preservable) molluscan debris ought, in general, to be significantly older than foraminiferal assemblages of the same horizon. Durations of time averaging should therefore be greater for molluscan assemblages. Foraminifera at Choya Bay undergo seasonal reproduction caused by periodic overturn (late March–April, late fall) of the nutrient-rich thermocline and phytoplankton blooms (Martin et al., 1996a). Unlike *Chione*, however, most foraminifera at Choya Bay dissolve within 3 mo following reproduction. Rapid dissolution of foraminiferal tests is not surprising, given their small size (and thus likely chemical reactivity) and the low total shell content near the SWI at most sites at Choya Bay. Similar reproduction–dissolution cycles have been widely reported for foraminifera from other shallow-water environments (Green et al., 1993b; see Martin, 1993, for further references), and Powell et al. (1984) estimated half-lives of 100 days for the smallest (0.8–3.1 mm) juveniles of molluscan death assemblages in Laguna Madre, Texas. Kowalewski (1996) also noted differential loss between bivalves and lingulide brachiopods: Bivalves persist for at least 100 years, whereas brachiopod valves are destroyed within a matter of months.

Despite the rapid disappearance of much of the seasonal reproductive pulse, foraminifera from the northern margin of Choya Bay were as old as or older than the much larger *Chione*. Martin et al. (1996a, 1996b) suggested several mechanisms to explain their results. First, tests at northern flat sites are better preserved, and therefore older, because of higher total shell content near the surface and lower rates of bioturbation (alkalinity buildup). Calculated sediment accumulation rates at Choya Bay for the past few thousand years are quite low (about 0.038 cm/yr; Flessa et al. 1993) because sediments are de-

rived mainly from local sources. Without high sediment accumulation rates, conveyor belt deposit feeders (CDFs, primarily callianassid
shrimp and polychaetes) repeatedly transport fine-grained sediment
from depth up to the SWI; at the same time, this tends to concentrate
coarse molluscan (mainly bivalve) debris in a distinct subsurface shell
layer (biogenic graded bedding). The depth to the shell layer decreases
northward from over 60 cm on the southern flat to about 10 cm in
some places, especially over a Pleistocene coquina that is about
125,000 years old (antecedent topography). Consequently, total carbonate weight percentage (shell content) of sediments near the SWI
tends to increase to the north (Martin et al., 1996a).

The activities of CDFs also influence pore water chemistry (Martin
et al., 1996a), and are most intense on the inner and southern flat and
decrease to the north, especially where sediment thickness is below
20 cm (Martin et al., 1996a). Sediment at these sites tends to become
anoxic, and sulfate-reducing bacteria, which are oxyphobic, appear to
thrive at northern flat sites, as evidenced by pore water alkalinity as
high as 50 meq)/l (normal seawater alkalinity is about 2.5 meq/l; in
this respect, the interested reader should compare the FOAM and
NWC sites of Long Island Sound discussed in Canfield and Raiswell,
1991).

Two other mechanisms may account for the anomalously old ages
of Choya Bay foraminiferal assemblages. One involves the upward reworking of significantly older tests via bioturbation or from erosion
by storms from the underlying Pleistocene limestone that rims the
northern margin of Choya Bay. The other also involves sediment mixing: Despite apparent extensive dissolution of foraminifera following
reproduction at Choya Bay, some tests are probably rapidly incorporated into the subsurface shell layer by CDFs and preserved there until
they are exhumed, much later, by biological activity or storms. Thus
even if a shell is rapidly buried by downward advection by CDFs at
Choya Bay, rather than by rapid sediment influx, it may still remain
pristine. The third mechanism is essentially a compromise between
the first two mechanisms, as time averaging is certainly plausible on
shorter timescales than those of the second mechanism.

None of these mechanisms is mutually exclusive, although the
third mechanism seems the most plausible. With respect to the first
mechanism, higher shell content and alkalinity near the surface at
northern flat sites may indeed slow rates of shell dissolution following
reproductive pulses, but it seems unlikely that this mechanism alone
can account for the surprisingly old Choya Bay test ages considering

the extent of presumed test dissolution in these sediments (Martin et al., 1996a). With respect to the second mechanism, the potential for mixing or reworking of substantially older (Pleistocene) foraminifera into Holocene sediments at Choya Bay was indicated by the high A/I values of certain *Chione* valves mentioned previously. Such specimens that had been potentially reworked from the underlying Pleistocene were noted only rarely in our samples, and they were not used in ^{14}C analyses.

Indeed, reworking of substantially older (hundreds of thousands to millions of years) or younger microfossils into younger (older) sediments by upward leaking or vice versa by piping is not usually considered a serious problem by biostratigraphers. In most cases, biostratigraphic zonations are sufficiently precise at these timescales that reworked specimens (including, presumably, microfossils) are typically recognized by their anomalous stratigraphic occurrence and qualitative state of preservation (surface roughening, breakage, infilling with foreign sediment). Berger and Heath (1968) concluded that bioturbation would not normally be responsible for such reworking, as the concentration of the reworked marker would probably be on the order of 10^{-6} of its original value when the thickness of sediment separating recent from older material was as little as 1 m (cf. figure 4.7). They concluded that erosional processes were much more important in reworking specimens. Cutler and Flessa (1990) came to a similar conclusion: Bioturbation is an inefficient means of producing stratigraphic disorder, but physical reworking disorders sequences rapidly.

Nevertheless, extensive mixing or reworking of old microfossils with little or no surface alteration into much younger sediments may be more widespread than previously thought. For example, based on amino acid racemization techniques, Murray-Wallace and Belperio (1995) found that apparently fresh, uncemented specimens of the large, symbiont-bearing foraminifer *Marginopora vertebralis* were reworked from late Pleistocene (about 125 ka) carbonate sediments into the overlying modern carbonate skeletal sands (see also Wehmiller et al., 1995).

THE STRATIGRAPHY OF SHELL CONCENTRATIONS

Consider what happens to the stratigraphic distribution of fossils when sediment accumulation rates change as environments shift, and how this affects the information content of fossiliferous strata. Kidwell (1988, 1989, 1991, 1993b) has documented the formation of shell beds in response to shifting environments (especially changes in sea level) and corresponding changes in sediment accumulation rate. She first studied the Miocene shell beds found in classic exposures of the Calvert Cliffs region along the Chesapeake Bay and then tested her models of shell accumulation and taphonomic criteria (Kidwell, 1986; Kidwell and Bosence, 1991; cf. figures 4.1, 4.2, and 4.3) in estimating the relative magnitude of sedimentary hiatuses in other depositional settings. In low-subsidence (low-sediment-accumulation) settings, hiatuses of about 1 million years' duration were demarcated by either bare unconformities (sequence boundaries) or erosional surfaces covered with complex concentrations of shells (figures 4.8 and 4.9). These concentrations were of two types: lag deposits, in which case resistant hardparts had been exhumed and concentrated from underlying (older) deposits during significant erosion and stratigraphic truncation; and hiatal concentrations, in which numerous generations of shells accumulated in a single, thin deposit (compared to those of coeval or "equal age" strata) while sediment accumulation remained low (but higher than for lag concentrations). In higher-subsidence (higher-accumulation-rate) settings, sequence boundaries were represented by either composite concentrations (which are similar to hiatal concentrations but in which the total thickness of the shell bed surpasses that of coeval strata) or event concentrations (which represent single, ecologically brief episodes of shell accumulation).

Thus, as emphasized in chapter 2, certain depositional environments may be more suitable than others, depending on the questions being asked. In Kidwell's studies, settings with low sediment accumulation rates along the passive margin of the eastern United States (Miocene Calvert Cliffs) were characterized by lag and hiatal concentrations that are no doubt suited for fairly coarse biostratigraphic and paleobiological studies. On the other hand, high accumulation rate settings located along active plate margins (such as those of the West Coast of the United States) were characterized by composite and event concentrations that would be most useful in bridging the gap between ecological and evolutionary processes (chapter 5).

To date, most actualistic studies (investigations of modern settings that may serve to unravel ancient ones) of taphonomy and time averaging have concentrated on nearshore environments. These settings are easily accessible and play an important economic role (such as salt marshes as fisheries and filters of pollutants and inland bays and coral reefs for recreational activities such as sailing and scuba diving).

The situation is more nebulous for assemblages formed in deeper water. Although shelf and slope sediments have been sampled extensively by petroleum companies for many years, especially for microfossils because these are easily obtained in large quantities from well samples, these companies have emphasized coarse biostratigraphy and bathymetry. Consequently, we know a lot about the distribution of microfossils, but little about how their assemblages form.

FIGURE 4.8 Formation of shell concentrations in relation to sediment accumulation rate. Compare with figure 4.9. See text for further discussion. *From S. M. Kidwell in* Taphonomy: Releasing the Data Locked in the Fossil Record, *edited by P. A. Allison and D. E. G. Briggs. © 1991 by Plenum Press. Used with permission of the author.*

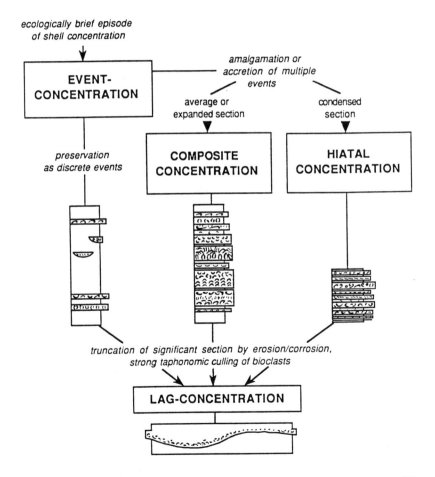

FIGURE 4.9 Shell concentrations in relation to degree of time averaging. *From S. M. Kidwell, 1993a. Time-averaging and temporal resolution in Recent marine shelly faunas. In S. M. Kidwell and A. K. Behrensmeyer, eds., Taphonomic Approaches to Time Resolution in Fossil Assemblages. Paleontological Society Short Courses in Paleontology No. 6:9–33. Reprinted with permission of the Paleontological Society.*

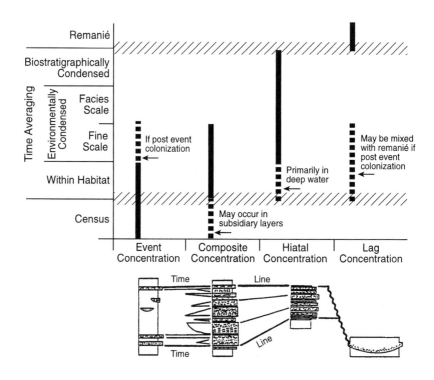

TAPHONOMIC GRADIENTS IN DEEPER SHELF AND SLOPE SETTINGS

Based on a literature review, Flessa (1993) estimated durations of time averaging of bivalve assemblages of up to 10,000 years for deeper shelf settings. Whether offshore microfossil assemblages exhibit similar degrees of time averaging remains to be determined.

Martin (1993) speculated on a series of taphonomic gradients across the continental shelf and slope. Like nearshore settings, ichnofacies models across continental margins (figure 4.10) reflect factors such as wave energy, food (organic carbon) availability (Berger and Killingley,

1982), oxygen content, and sedimentation rate. All of these factors tend to decline away from shore, so that burrowing is increasingly restricted to near the SWI. Lin and Morse (1991), for example, concluded that in the northern Gulf of Mexico (Mississippi Delta and Texas–Louisiana shelf), SO_4^{2-} reduction rates and pyrite (FeS) concentrations generally decrease exponentially with increasing water depth, and reflect decreasing sedimentation rates. Sulfate reduction rates in this region are intermediate between those of shallow nearshore organic carbon-rich sediments and organic-poor deep-sea sediments (Canfield, 1991; anoxic silled basins are an exception; see Berger and Soutar, 1970). Organic carbon concentrations in sediment are also a function of sedimentation rate: At high sedimentation rates, labile organic matter undergoes a shorter period of oxic and suboxic degradation before

FIGURE 4.10 Distribution of trace fossils associated with a transect across the continental margin. Vertical traces dominate nearshore, whereas more complex, horizontal traces characterize deep water. *From A. A. Ekdale et al., Ichnology: the use of trace fossils in sedimentology and stratigraphy. Society of Economic Paleontologists and Mineralogists Short Course Number 15, Tulsa, Okla. Reprinted with permission of SEPM (Society for Sedimentary Geology).*

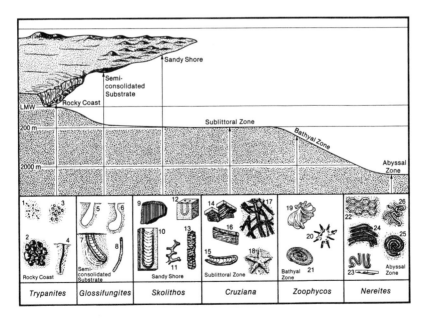

it is incorporated into subsurface layers where it can support $SO_4^=$ reduction and raise alkalinity (Canfield, 1991; Lin and Morse, 1991), hence the old saw that good hardpart preservation depends on rapid burial (regardless of $CaCO_3$ content).

The commonly observed peak in foraminiferal (benthic and planktic) abundance near the shelf–slope break (see Bandy, 1953) is probably a reflection of surface water productivity (shell input, as evidenced by increased planktic/benthic ratios and foraminiferal numbers) and decreased rates of terrigenous sedimentation. As $SO_4^=$ reduction rates decline across the continental shelf (Lin and Morse, 1991), rates of bioturbation must also either decline or be counteracted by increased input of $CaCO_3$ for preservation to occur. Across the shelf break and onto the slope, $SO_4^=$ reduction rates presumably continue to decrease, which presumably reflects decreased surface water productivity away from shore and increased oxic degradation of organic carbon. Indeed, the lysocline tends to shallow toward the continental margins because of the decay of organic matter derived from land and plankton (Berger, 1977). Thus there is a tradeoff between the rate of bioturbation, which is typically highest in nearshore areas and decreases across the continental margin (note D values in Matisoff, 1982) and is dependent on organic matter concentration (food), and the rate of SO_4^{2-} reduction, which is also dependent on organic carbon influx but is inhibited by bioturbation (which oxygenates pore waters). All three (bioturbation, SO_4^{2-}, and carbon influx) influence preservation, but all are related to sedimentation rate.

Just as for shallow-shelf environments, temporal resolution of deeper-shelf settings must be assessed, taking into account each environment's taphonomic peculiarities. In the case of sediments located downslope from large deltas, for example, increased shell and organic matter input (a consequence of nutrient influx), coupled with rapid burial, probably elevate SO_4^{2-} reduction rates, enhance preservation, and maintain a fairly normal age structure. Rapid changes in deltaic accumulation (lobe switching) may significantly affect the continuity of sedimentation, however (see chapter 2).

On the lower slope and rise, as pelagic rain (productivity) rates tend to decrease and a localized lysocline may be approached, differential preservation may again become important, especially if relevant biostratigraphic or paleodepth indicators are rare or susceptible to dissolution. Benthic foraminifera, for example, are typically more resistant to dissolution than planktics and the benthic/planktic ratio has often been used as an index of postmortem alteration of deep-sea assemblages (Thunell, 1976).

Nevertheless, differential preservation of benthic foraminifera also merits further study. Douglas et al. (1980) concluded that life-and-death assemblages of the upper slope (100–400 m) of the southern California borderland are strongly similar (based on surface samples), but Loubere and Gary (1990; see also Loubere et al., 1993a, 1993b) concluded that there was substantial specimen loss in the upper 10 cm of boxcores from the slope (about 1000–1100 m) off the Mississippi Delta. Denne and Sen Gupta (1989) came to similar conclusions for Texas–Louisiana slope sediments. They found that bathyal calcareous species could be grouped into those that are well preserved (increasing downcore), moderately preserved (no trend), and poorly preserved (decreasing downcore). Agglutinated species were either disaggregation resistant (no abundance trend downcore) or disaggregation prone (decreasing downcore).

"LAWS" OF TAPHONOMY?

Although taphonomic processes have been studied for many years, they have received the intense scrutiny they deserve only in the last decade or so, as paleobiologists have come to focus on the potential record of processes that occur on ecological and short geological timescales. Coarser stratigraphic resolution is not good enough, especially as taphonomic losses may artificially shorten or extend the range of stratigraphic markers (chapter 2).

Are there some "laws" that might guide us in future efforts? Perhaps, but these "laws" are really rules because so many variables (and historical circumstances) enter into the formation of a fossil assemblage (Wilson, 1988, among many others, discusses this topic in greater detail). Some of these rules have already been mentioned:

- In most cases, in order to have a reasonable chance of preservation, an organism must possess hardparts (Lagerstätten are an exception).
- Burial must be rapid so as to minimize disarticulation, dissolution, and other taphonomic processes (that is, transfer through the mixed layer or taphonomically active zone must be rapid).
- Burial in fine-grained sediment is preferred because fine-grained sediments are deposited in low-energy (quiet water) en-

vironments, which limits abrasion and breakage, and because the permeability of such sediments is low, thereby slowing dissolution in pore waters.

- Dissolution and bioerosion are probably more important than abrasion in destruction of marine shells, whereas transportation and disarticulation may be more important in the terrestrial realm.

Although Kidwell's studies of shell beds represent a synthesis that allows us to predict the information content of fossiliferous strata and test hypotheses, for the time being, more precise "laws" than these may lead us astray. Indeed, other intuitive rules (actually assumptions) suggest themselves, but as we have seen they are not necessarily correct:

- Big shells last longer than small shells.
- The condition of a shell is an indicator of its age.

Perhaps this is because each taphonomic setting is unique, depending on the historical circumstances that constrain any taphonomic "laws" (or, more properly, principles; Olson, 1980). Indeed, so far taphonomy has been characterized by its inductive nature (see the Prologue); it depends on empirical generalizations that are derived from the study of many modern and ancient settings (see chapter 3 of Lyman, 1994, for extensive discussion). More recent syntheses and models (such as taphofacies) based on numerous case studies are, for now, the best approximations that we have to rigorous taphonomic principles, and other models (such as those of Martin, 1993) require extensive testing. Given the surprises that await us (like those regarding shell age at Choya Bay), more case studies are needed to constrain and develop models of preservation. Nevertheless, this should not prevent us from examining the fossil record of ecological and evolutionary processes.

5

Biological Processes Inferred from the Fossil Record

Does ecology matter! —*J. B. C. Jackson (1988)*

Time flies like an arrow; fruit flies like a banana. —*Marx*

By now the reader must be convinced that nothing concrete can be obtained by studying the fossil record. Studying Earth's history must be a complete waste of time because the geologic record is too incomplete or too scrambled to document or prove anything. Nevertheless, the fossil record not only allows us to assess the impact of ecological processes over periods of time much longer than those normally considered by an ecologist, but also reveals that there are likely to be processes (as indicated by patterns in the fossil record) that occur only on long timescales. To corrupt Marshall McLuhan's famous phrase, the pattern is the process (maybe). Moreover, time-averaging of fossil assemblages actually *enhances* expression of ecological signals because it damps (filters) out short-term noise (Wilson, 1988; McKinney, 1991). That is not to say that understanding how ecological signals are filtered or smoothed is not important; it is if we are to understand what gets incorporated into the fossil record (hence all the equations in chapter 3).

These patterns, and presumably the processes that produced them, are by no means strictly academic. They bear strongly on humankind's impact on the health and biotic diversity of the planet. Paleontology and stratigraphy bear on such environmental phenomena as population dynamics, speciation, and extinction; the organization and resilience of biological communities to disturbance; and the occurrence of

alternative community states in response to environmental disturbance, both natural and anthropogenic.

Besides attempting to answer these broad questions (and posing
new questions in their stead), paleontology and stratigraphy continue
to affect us directly on the short timescales of our everyday lives. Ironically, we have the means to answer questions about the state of the
planet's health, but we have historically concentrated on finding resources that will alter it (oil and gas), and we will continue to do so as
long as civilization remains dependent on fossil fuels. On the other
hand, we can also answer such questions as those about the state of
the environment before the spread of humans to an area. What constitutes a pristine environment? How does this affect legislation de-
signed to preserve (or destroy) wetlands, for example? What have been
the effects of land use and runoff (including fertilizers, sewage, and
nutrients released from deforested soils) on natural waters over the
past few decades or the past two centuries? How fast is sea level rising?

UPWARD SCALING OF ECOLOGICAL PROCESSES?

Before proceeding, we must again consider the issue of scale. Upward
scaling (extrapolation or reification, in the parlance of Allen and Starr,
1982) of ecological processes to geological (evolutionary) timescales is
a principal concern in interpreting the fossil record. Aronson (1994)
concluded that many biological processes are scale independent and
suggested several examples in which ecological (short-term) processes
scaled upward to much longer (geological) scales. He suggested, for
example, that the trend of increasing development from temperate latitudes to the tropics of gastropod-crushing morphologies in bony fish
and certain crustaceans and defensive morphologies found in their
prey (such as heavy calcification, tightly coiled shells, spines, and ribs)
scaled upward to the Mesozoic marine revolution documented by
Vermeij (1987), during which a trend of increasing predation and
predation-related morphologies first became evident in the fossil record; this trend continued to the present (possible driving forces for
the trend are discussed in chapter 7). In another example, he suggests
that the absence of abundant brittle stars (ophiuroids) in modern marine environments (small spatial and short temporal scales) is indicative of increased predation; fossil brittle star assemblages also decrease
in frequency beginning in the Mesozoic, as the marine revolution got
underway ("geological" scales of time and space). Other workers have
also documented ecological processes in the fossil record, such as pre-

dation and shell use by hermit crabs, even in environments highly unfavorable to preservation (see Kelley and Hansen, 1993; Walker and Carlton, 1995).

According to Aronson (1994), the origin and development of the three major evolutionary faunas of the Phanerozoic (figure 5.1) represent upward scaling from ecological processes (see also Sepkoski, 1992). Each of these faunas appeared near shore and moved offshore through geologic time, replacing the preceding fauna (Jablonski and Bottjer, 1989). Aronson (1994) argues (via extrapolation from ecological scales) that this is also a matter of increasing predation intensity, as evidenced by the effects of overfishing by humans in nearshore waters (I suggest additional reasons for this pattern in chapter 7). Darwin (1859) drew similar conclusions based on competition: earlier faunas or floras would be exterminated by later ones.

Thus it appears that in many cases ecological processes do scale upward. However, as Aronson (1994:448) argues, different processes also operate at different scales (table 5.1). He suggests that although small-scale processes may differ from those that act over much larger spatial or longer temporal scales, the basic response of the biota is the same. That may often be the case, but it begs the question of what the processes are (table 5.1). Until more systems have been examined, upward scaling in each situation must be evaluated on its own merits, as discussed later in this chapter.

POPULATION DYNAMICS AND EXTINCTION IN THE FOSSIL RECORD

As mentioned previously, species that are geographically widespread, and therefore likely to be eurytopic, are more likely to survive extinction than species that are stenotopic (narrowly tolerant of environmental change). Stenotopic species are more likely to be geographically restricted, and their populations may be smaller and undergo greater fluctuations in abundance than do those of eurytopic species (Jablonski, 1989; Stanley, 1990a). Eurytopy is also more likely to promote a wider variety of trophic (feeding) and other sorts of biotic interactions than stenotopy (Futuyama and Moreno, 1988; McKinney et al., 1996).

McKinney and Frederick (1992) tested this hypothesis using the abundances of 25 species of benthic foraminifera collected at 1-m intervals from a single section of Eo-Oligocene limestone in Florida. They analyzed the population abundance fluctuations through time

FIGURE 5.1 The major evolutionary faunas of the Phanerozoic delineated by a statistical technique known as factor analysis. Boundaries are intergradational. Note movement through time from nearshore to offshore. Major taxonomic components of each fauna are indicated. *Modified from Sepkoski and Miller, 1985.*

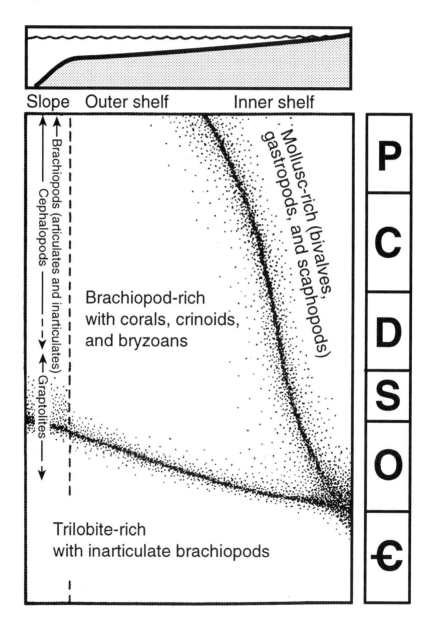

TABLE 5.1 Hierarchy of Community Dynamics at Different Temporal Scales

Process	Preserved Pattern	Estimated Duration (years)	Temporal Dynamics Involved
Community evolution	Community lineages	10^5–10^7 (?)	Origination and elaboration of new types of communities; community structural divergences; major biogeographic displacements and turnovers caused by extinctions (both species–environment and species–species interactions involved)
Community replacement	Community sequences	10^2–10^6 (?)	Abrupt to gradual community transitions caused by environmental changes; long-term population–environment interactions predominate
Patch development	Amalgamated seres or pseudoseres	10–10^3 (?)	Prolonged occupation of seafloor area by benthic community undergoing many episodes of succession or response
Ecologic succession	Seres	1–10	Autogenic, biotic changes in community composition and structure; short-term organism–organism interactions predominate
Community response	Pseudoseres	1–10	Allogenic, successionlike community changes; seasonal or cyclic short-term responses to environmental changes not leading to replacement; organism–environment interactions predominate
Community establishment	Noninteractive colonization sequences	10^{-3}–10^{-1}	Earliest stages of substrate invasion in newly opened habitats; strongly dependent on availability/proclivity of colonizing larvae and motile adults; organism–environment interactions predominate

Durations are approximate. Modified from Miller, 1986.

(upsection) using a fractal-based technique called rescaled-range analysis (Sugihara and May, 1990). Despite the short-term incompleteness of the rock record, the method works because it is fractal based (independent of scale). Let $n(t)$ equal the abundance of a species during a particular time interval t, and let $n'(t)$ represent the normalized deviations of abundance

$$n'(t) = \bar{n}(t) - n(t) \tag{5.1}$$

for each t between 0 and T (the entire time interval). The rescaled range is then

$$R(T) = \max n'(t) - \min n'(t), \tag{5.2}$$

where $0 \le t \le T$. Variations in $R(T)$ through time are then analyzed with a fractal model

$$R(T) = cT^H, \tag{5.3}$$

where c is a proportionality constant $[R(T)\alpha T^H]$ and $H = 1/F$, where F = the fractal dimension (see chapter 1). Remember that $F = 1$ for a straight line and $F = 2$ for a plane. If $F = 2$, then $H = 0.5$, which means that the plane is "filled" by the curve and represents a random walk (Sugihara and May, 1990), and the time series (the collection of all n over the entire time interval T) of rescaled abundance variations R represents a random walk (Brownian motion) no matter what the scale of Δt (duration of observation). Thus abundance fluctuations show no persistent trend in a particular direction (increase or decrease).

As F decreases from 2 to 1, H increases above 0.5 and R becomes less plane-filling (tending toward a line). In this case, there is a persistent trend of increasing variation in abundance at every scale of Δt. There will be higher highs and lower lows because as Δt increases, variation also increases. On the other hand, if H decreases to below 0.5, variation decreases ("antipersistent").

McKinney and Frederick (1992) calculated H for the abundance time series of each species of foraminifer. H was calculated from the slope of the line in the plot of log $R(T)$ versus log T because the logarithm of $[R(T) = cT^H]$ equals log $R(T) = H$ log T + log c (in which H = slope of the line $y = mx + b$, where x = log T and b = log c). They then compared the H values of each species to the stratigraphic range of the species in the fossil record to see whether species with high (low) H values were indeed extant for short (long) intervals of geologic time. They found a statistically significant negative correlation ($r = -0.60$; $p < 0.05$) between H of a species and its stratigraphic range; that is, as H decreased, stratigraphic range increased. In other words, populations with greater abundance fluctuations go extinct sooner than those that maintain more stable populations. (The fact that the absolute value of r or the correlation coefficient equals 0.60 means that about 36 percent of the variation, or the degree to which the data points fall off the line, is accounted for by the plot of H versus range because r^2 = total amount of variation; p is the alpha error, or the probability that this relationship could have arisen by chance alone. In this case, $p < 0.05$ indicates that the relationship could have arisen only 5 times out of 100 by chance.)

Note that McKinney and Frederick (1992) are dealing with a single section and therefore *local* populations. Whether this relationship holds on larger (such as regional) scales remains to be seen, but available data seem to suggest that it does (see Stanley, 1990a, for the fossil record; Harrison, 1996; Maurer and Nott, 1996, for modern populations) and that, in this case, ecological processes appear to scale upward.

Metapopulation theory (which has garnered a great deal of attention from ecologists and some paleobiologists recently) states that a species with many small, ephemeral populations can persist *longer* than a species represented by a few large populations. According to this theory, the distribution of a species can be modeled as a mosaic of populations "winking on and off" like Christmas tree lights, representing local colonization (immigration) and extinction, respectively. Does this theory of local population extinction on ecological spatiotemporal scales extrapolate upward to extinction on geologic scales? Probably not. First, it is unlikely that the metapopulation model applies to most real populations in ecological time, much less geological time, because the geographic distribution of real species is much more complicated than metapopulation theory assumes, consisting of large, persistent source populations that supply small, ephemeral sink populations (Harrison, 1996; Maurer and Nott, 1996; McKinney et al., 1996). Second, species can persist because they can track environmental change (Thomas, 1994; Holland, 1995; McKinney et al., 1996; see also chapter 2). Third, as demonstrated by the fossil record, rare species are less likely to persist than common ones (see also later discussion of the particle model).

101

EVOLUTION, HIERARCHY, AND THE PRINCIPLE OF COMPLEMENTARITY

Ever since Darwin, there has been considerable debate about the exact mechanisms of evolution. The pendulum of thought on this matter has certainly swung back and forth over the last hundred or so years, and debate has often been heated between evolutionary biologists and paleobiologists. This is not a debate among evolutionists about whether biological evolution is a fact (contrary to what creationists wish us to believe), but a debate about *how* evolution proceeds.

In his recent review of this chasm between allies, Eldredge (1995) makes a strong case that there are evolutionary processes that act over scales of time much longer than those considered by most population geneticists (with the exception of Theodosius Dobzhansky), and that are only recognizable from the fossil record. On the other side are most biologists (reductionists or ultra-Darwinians, in Eldredge's parlance), who, like Darwin in *The Origin of Species*, believe that *all* evolutionary change is a result of alterations (mutations) to the genome (DNA) of organisms that then slowly spread through populations and transform them into new species. Organisms that are more genetically fit are selected (survive) and are therefore more likely to leave offspring bearing the same genetic traits (differential reproduction). According to the ultra-Darwinians, these lower-level processes scale upward to account for the evolution of higher taxa (species). Indeed, Burlando (1993) concluded that taxonomic patterns—and therefore evolutionary processes—are fractal (self-similar), consisting of many isolated lineages and clumps, the clumps in turn consisting of isolated lines and clumps, and so on.

Thus ultra-Darwinians view all evolution as slow and gradual (phyletic gradualism). Eldredge (1995) argues, however, that the fossil record indicates periods of mostly slow, nondirectional change (stasis), punctuated by abrupt speciation. (See also Eldredge and Gould's 1972 paper on punctuated equilibria, which brought evolutionary biology out of the torpor of the "modern synthesis" of the 1950s and 1960s. Eldredge (1995) discusses how most evolutionists—biologists and paleontologists alike—came to view evolution as a slow, gradual process, after having earlier advocated that evolutionary rates could vary widely.)

Eldredge and Gould's (1972) model of punctuated equilibria is really the record of allopatric speciation preserved in the fossil record (figure

102

5.2). The allopatric model, which holds sway among most biologists (although there is increasing evidence that sympatric speciation, in which species ranges overlap, occurs; Gibbons, 1996), was originally formulated by ornithologist Ernst Mayr in 1942. Basically, Mayr (1942) said that new species originate by geographic isolation of local populations (demes) by rivers and streams, mountain chains, or simply changes in local climate (sunlight, moisture) from, say, one side of a valley to another. Genetic transformation of the demes then occurs through either the founder effect, in which the founding (or newly isolated) population is not genetically representative of the original parent population from which it came (in statistics, this is akin to taking a sample from one of the extremes of a bell curve); or genetic

FIGURE 5.2 Phyletic gradualism and punctuated equilibria. Rates of evolution are represented by slopes of the branches. (a) Phyletic gradualism, in which species change gradually through time; in this case, there is a net shift toward longer leg length through time. (b) Punctuated equilibria, in which new species arise rapidly and are followed by long intervals of little or no net evolutionary change; in this case, 15 of the new 16 species have longer legs than their predecessors. (c) Punctuated equilibria plus species selection (sorting). In this case, equal numbers of species arise with either shorter or longer legs, but the net trend is toward greater length. The trend may result from species with longer legs tending to survive longer and leaving more descendant species or when species with longer legs tend to produce new species at higher rates. By analogy with natural selection of individuals, species are selected, although Eldredge (1995) prefers the term *sorting* because the selection of individuals and species may actually result from different processes. *From* Earth and Life Through Time *by S. M. Stanley. Copyright © 1989 by W. H. Freeman and Company. Used with permission.*

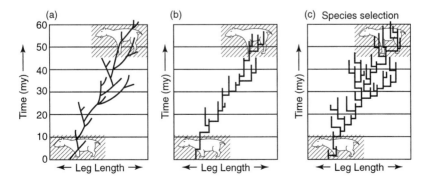

103

drift of gene frequencies within a local population. Random changes or mutations occur in DNA base triplets (codons) that code for amino acids in proteins such as enzymes, which catalyze biochemical reactions and are highly specific to the reactions. These changes then alter ecological traits (such as food and habitat preferences and mating behavior) and promote reproductive isolation and speciation (members of a species interbreed with one another to produce fertile offspring). Genetic drift is a kind of "sampling error" in which some genetic traits are accidentally lost from small populations (such as demes) whereas others are passed on (Salthe, 1985). As beneficial mutations spread through the population, the organisms eventually become reproductively isolated from members of other populations (including those of the original parent population), and new species emerge, members of which do not interbreed and produce fertile offspring with members of other species. On the other hand, evolutionary stasis, which, according to Eldredge (1995), occurs between episodes of speciation, may simply reflect the canceling out of minor, random genetic changes within demes (Eldredge, 1995). This pattern conforms precisely to a hierarchical model. Changes in gene frequencies produce the lower-level initiating conditions, which are constrained by upper-level factors such as environmental patchiness that separate demes (see chapter 1).

However, according to Eldredge (1995), the patterns in the fossil record tell us that the entire evolutionary process is much more complex than even this. Macroevolutionary processes occur at higher hierarchical levels (evolution above the species level) that differ from those at the lower levels of microevolution (changes of gene frequencies within demes). Trends or directions within evolutionary lineages, for example, may be caused by a process called species sorting, which acts at a higher level than does natural selection. Species selection or, as Eldredge (1995) prefers, species sorting (to distinguish it from the process of natural selection) is *not* merely an upward scaling of natural selection acting on species. In fact, speciation itself is *not* adaptive. Species sorting is not directly connected with mutations in DNA; rather, it is a process of pruning. Some species are more likely to leave "offspring" (new species) than others (differential births and deaths of species; figure 5.2). Specialized species are more likely to undergo speciation (and extinction) than generalized ones (see chapter 2; Stanley, 1979, makes a similar argument). According to Eldredge (1995), among eurytopic species (meaning that the organisms that make up the species are eurytopic) there is such broad niche overlap between the ancestral species and their descendants that the newly evolved fledgling species must diverge rapidly from the parents in order to persist as separate species. On the other hand, fledgling species that have evolved from narrowly adapted (stenotopic) parental lineages may have an easier go of it in exploiting available niches and surviving because they must diverge less from the parent lineages.

Both Eldredge (1995) and Vrba (1984) view species as being sorted according to their emergent properties, that is, properties that are more than the mere sum of the parts (Allen and Starr, 1982:38–39; Cracraft, 1985). A clear view of the parts may actually preclude observation of an emergent property. For example, if we focus on an atom (or a few players in a football game; see chapter 1), we may not see the atom's bonds with other atoms (or the players' interactions with other participants) until we back up and, so to speak, see the forest despite the trees or the molecule despite the atoms. Thus a higher-level property (species) not seen at lower levels of a biological hierarchy emerges because of a change in the way observations are made (measured), as long-term patterns of species origination and extinction are observed in the fossil record.

Unfortunately, species sorting was originally called species selection, which has caused a great deal of rancor because it may be misinterpreted as saying that the *process* of natural selection also occurs at higher hierarchical levels, when only a *pattern* has been observed in the fossil record, not necessarily a process. Ultra-Darwinians insist that mutation and adaptation benefit the organism, not a population of organisms or even an entire species (group selection). In fact, some evolutionary biologists think in such a reductionist way that they would deny the existence of species altogether (Eldredge, 1995).

But the question remains: Is species sorting merely a *pattern* in the fossil record or is it a *process* truly different from natural selection? Vrba (1984) suggested that emergent properties that would cause differential changes in speciation and extinction rates of a species, and that would be sorted at the species level, include characteristic population size, spatial and genetic separation of populations, and the nature of the species' "periphery" (whether smooth or complex). Without arguing against the action of different evolutionary processes at different hierarchical levels, Cracraft (1985) argued that emergent properties such as those just described are not heritable from species to species because they are all extrinsic to the species, are probably not even constant over timescales shorter than those required for speciation to occur, and depend on "historical contingencies [that] do not

105

repeat themselves" (p. 226; we return to this topic again in chapter 8).

This apparent conflict between evolutionary biologists and paleobiologists who also study evolution in the fossil record illustrates what Allen and Starr (1982) call the principle of complementarity: Paradoxically, the parts of unified models sometimes appear to contradict one another. Allen and Starr (1982) are not talking about a tolerance of alternative models by different workers, but the integration of seemingly contradictory models. The principle of complementarity recognizes that "incompatibilities are necessary between dual modes of description, both of which are required for a complete account of phenomena. . . . The system under observation may be unified, but the requirements for description are dual" (Allen and Starr, 1982:43; see also chapter 5 of Allen and Starr, 1982, from which the following discussion is taken).

The dynamics of lower levels are rate dependent because they depend on *laws*, which are "inexorable . . . incorporeal . . . and universal," but they are constrained by *rules* at higher levels, which are independent of lower-level rates because they are "arbitrary, . . . structure-dependent, [and] local. In other words, we can never alter or evade laws of nature; we can always evade and change rules" (Allen and Starr, 1982:42). This is precisely what Eldredge (1995) has argued in contrasting Mendel's laws of segregation and independent assortment of "genetic factors" (that were recognized by him based on statistical analysis alone and were only later correlated with the chromosomes that bear genes) with species sorting. Mendelian genetics acts at low levels and is rate dependent, whereas species sorting is not a law but a rule, acting not invariably at higher levels and independent of lower levels (such as genes). However, most biologists have taken a highly reductionist (mechanistic) approach. "There is a temptation to reify natural selection and to seek the level where selection is 'really' occurring. . . . Our hierarchical view requires selection at several levels simultaneously. . . . In evolution . . . events at the lower level have their significance concentrated at the higher level with the passage of time. Organisms of favored structure saturate the next higher level, the population, so that it develops new ordered characteristics [the same could be said of genes and organisms or populations and species]. . . . The integrative tendencies of holons are not selected principally at the level at which these tendencies are manifested, but rather they contribute to the stability of the system of which they are a part and so are selected there. . . . The integrative tendencies of the holon contribute significantly to the selection of the next higher stable level and therefore to the evolution of the second higher stable level. . . . We only take exception to an obligate reductionism to the level of the individual organism or the single gene. With any hierarchy, signals of different strength reach a given holon from different levels [cf. figure 1.6]. It is not surprising that individual selection is the strongest evolutionary signal that we perceive, for as individuals ourselves we give heavy weight to that level of organization. Nevertheless, other selective agents at other levels will contribute" (composited from Allen and Starr, 1982:51–53, 100).

In his book, Stanley Salthe (1985) goes to great pains to describe a strict hierarchical view of evolution. To him, natural selection is a lawlike process, and to focus on the *process* itself is to take a "radically nonhierarchical [that is, reductionist] stance" (p. 174). In its most extreme and parsimonious form, the ultra-Darwinian approach reifies the gene level to all other levels; that is, selection at the gene level is transitive across all levels. Although this is seemingly the easiest and most logical approach (because it reduces the whole system to essentially a single level), he views this approach as at best a solipsism. In Salthe's view it means that "we simply [do] not obtain the fullest possible appreciation of the complexities of the real world" (p. 265).

According to Salthe, natural selection acts on populations, not genes or organisms. How could a gene or organism evolve when it has no significant extension in time? (Remember that one of the criteria for the recognition of objects or entities is spatio-temporal continuity; see chapter 1). Natural selection alters the gene frequencies in populations, or more correctly, the local breeding populations or demes, thereby causing the demes, which are isolated from one another, to climb local "adaptive peaks." The peaks are of slightly different heights, though, because no two demes are exposed to the exact same selective pressures. Now suppose that an entire area and its demes are exposed to a catastrophic fluctuation, such as an extreme flood, and that the adaptive peaks are differentially susceptible to extinction by flooding, so that some demes survive flooding. According to Salthe, natural selection is responsible for climbing the adaptive peaks, but *interdemic selection* operates during the floods. Thus the products of natural selection are regulated by a higher-level process that acts less frequently (Salthe, 1985:255–256).

The standard ultra- (or neo-) Darwinian approach, on the other hand, would view flooding as a general, average threat to the *entire* population that caused certain demes to climb adaptive peaks higher than others, which accounts for their survival. Certainly, there is a distribution of event intensities; less intense events (floods) occur more often than big ones (we return to this point in chapter 8), so that the events impose a history on the populations. But demes (populations) do not "see" average environments. The ultra-Darwinian (nomothetic) approach denies, then, the idiographic uniqueness of the floods because it considers only an *average* event.

In Salthe's view, the probability of survival in the flood example depends on the initiating conditions of each deme. Given that this particular system has operated for some time, each adaptive peak has its own particular value with respect to flooding; demes are not all the same because each deme has been subjected to slightly different selective pressures (including floods), genetic drift, and *preadaptation* or "latent potential" (Gould, 1986). Preadaptation refers to "a character of an organism evolved as an adaptation for one set of environmental conditions [which] turns out by chance to allow . . . occupation of quite different niche space" (Salthe, 1985:302). The ponderous limbs of primitive crossopterygians (lobe-finned fish, including the living fossil, the coelacanth) were preadapted for a rapid transition to land as amphibians. Primitive light-sensing and thermoregulatory organs may have been preadapted to give rise to eyes and wings, respectively, in arthropods. In experimental studies of insects, small wings are used to regulate body temperature; as the wings increase in size, advantages for flight become dominant just as thermodynamic benefits level off (Gould, 1996).

In Salthe's view, because natural selection operates on populations over a span of at most a few generations, it cannot be responsible for macroevolution. Moreover, Salthe regards macroevolution not as a process, but as a record. It is a summing up of processes occurring at higher levels (like those described above; see also chapter 8), a kind of trajectory or ontogeny of a lineage that occurs above the level of population (deme). The processes of macroevolution regulate processes at lower levels, called microevolution or "the generation-by-generation action of natural selection at the demic level" (Salthe, 1985:272).

Notice that evolution consists of three basic parts: population-based selection and drift, speciation, and "a loose conglomeration of patterns and interpretations relating to macroevolution . . . [that] have no pattern of relationships among themselves and relate only in an *ad hoc* manner to the speciation and selection theories" (Salthe, 1985:188). But if we view each of these as different levels of processes within a hierarchy there may be no fundamental conflict between them (principle of complementarity).

CAVEATS

Having been on both sides of the coin, as a zoologist and paleontologist, I must admit that Salthe (1985) and Eldredge (1995) make a persuasive argument for evolutionary processes that occur over long timescales and that do *not* scale upward from the simple changes in gene frequencies within demes. (Perhaps I am convinced because I have spent the last 15 years calling myself a paleontologist, not a zoologist). That being said, I am nevertheless a little uneasy about interpreting patterns in the fossil record without first evaluating the record itself (see chapter 2). Eldredge and Gould's (1972) original data for punctuated equilibria came from shallow-water Devonian trilobites and Bermudan land snails, but some of the best evidence for speciation and rapid evolution presumably comes from deep-sea cores because they apparently (and optimistically) represent long intervals of uninterrupted sedimentation (Schiffelbein, 1984). But even in deep-sea cores that exhibit continuous and high sedimentation rates, there is severe attenuation of signals with periods shorter than a few thousand years (Schiffelbein, 1984; see also chapter 3). Moreover, based on graphic correlation (see chapter 2), MacLeod (1991) demonstrated that periods of apparently rapid evolution within Neogene (Miocene to Recent) planktic lineages of foraminifera occurred during major changes in paleoceanographic regime and sediment accumulation rate. As the accumulation rate slowed, for example, apparent rates of morphological change increased because more of the morphological change, which is used to measure rates of genetic divergence and speciation, occurred within slow (condensed) intervals of sedimentation (see also Dingus and Sadler, 1982). Kidwell (1986) has also suggested how apparent directional trends (including those of phyletic gradualism) in the evolution of ancient lineages may be the result of artifacts of sediment accumulation (figure 5.3). In the case of macrofossils, these artifacts may be recognizable from changes in abrasion and fragmentation of shells, and encrustation by epizoans and bioerosion, all of which should increase when shells are not being buried rapidly and lie exposed at the sediment–water interface (SWI) for long periods of time (Kidwell, 1986). As discussed in chapter 4, however, this is not necessarily the case for microfossils.

FIGURE 5.3 Apparent trends in shell morphology within a fossil lineage that result from a net change in sediment accumulation rate. In this example, variation in morphology increases, fragile shells are selectively destroyed because they reside longer at the sediment–water interface, and individuals better adapted to shell-gravel conditions all increase upsection. *Modified from Kidwell, 1986.*

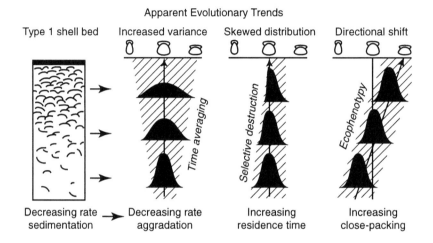

STASIS AND COMMUNITY UNITY?

What about change at the community level? Certainly the fossil record indicates significant change in ancient biotic communities through time (see chapter 8). Recently, some workers have suggested that a kind of "ecological locking" occurs, in which biological communities are resistant to environmental change below some threshold level (see DiMichele, 1994; Morris and Ivany, 1994; Lieberman et al., 1995). Using an approach similar to that developed by McKinney and Frederick (1992), McKinney (1996) and McKinney et al. (1996) concluded that much of the so-called stasis or locking is simply the result of measuring the most abundant (and presumably the most geographically widespread and eurytopic) species in the fossil record. "Entrenched" (eurytopic) species tend to recur together, are more likely to be sampled in the fossil record, are more likely to persist during times of environmental change, and may exhibit significant turnover only during major environmental crises. In contrast, rare species undergo much greater rates of turnover (speciation and extinction; Stanley, 1990a; McKinney et al., 1996) because they tend to be more stenotopic and suffer from greater population fragmentation ("edge effects"; Maurer and Nott, 1996). Although rare species typically far outnumber abundant ones (Koch, 1987; CoBabe and Allmon, 1994), rare species are more likely to be missing from stratigraphic sections because of destruction or lack of adequate sampling (McKinney, 1996; McKinney et al., 1996). The general effect of increasing bias against preservation and sampling would be to reduce the observability (and thus perception) of turnover in the fossil record (McKinney et al., 1996). Staff et al. (1986) documented changes in shell abundance and species dominance in living to modern-death assemblages in shallow-water molluscan assemblages of Texas bays, and Martin and Wright (1988) documented changes in foraminiferal abundances and diversity during the transition from life to death assemblages in modern backreef sediments off Key Largo, Florida.

McKinney and Allmon (1995), McKinney et al. (1996), and McKinney (1996) simulated the distribution of fossils (in this case, benthic foraminifera) in hypothetical sections using a particle model. This model specifies how widespread (and therefore abundant) species not only tend to recur in stratigraphic sections and therefore have longer stratigraphic ranges, but also co-occur (covary) as suites or assemblages preserved in the fossil record for long periods of time, whereas rarer species come and go. Rare species migrate into the area of deposition but are rarely preserved. Slightly more abundant species rarely or never migrate into the area and are never preserved in the section. Thus through both rarity and taphonomic loss, rare species are less abundant than common ones in a stratigraphic section (McKinney et al., 1996; McKinney, 1996). Although 80 percent or more of species in a region may be preserved (in this case the number of living molluscs in the Californian biogeographic province found in Pleistocene sediments of the same region; Valentine, 1989), it is unlikely that all 80 percent will be found in every sample from a stratigraphic section in this province (McKinney et al., 1996), and the high number of preservable rare species found in a sample must result from time averaging (chapter 4) of populations that exist and die over thousands of years (Peterson, 1977; Fürsich, 1978; Staff et al., 1986; McKinney et al., 1996).

Thus the perception of community stasis depends on the level of resolution (scale) used. Although community stasis probably does occur in the fossil record, it is measurable only for the most common species (McKinney et al., 1996). Like sediment-accumulation rates (Sadler, 1981), the degree of stasis increases with the duration of the interval examined (McKinney et al., 1996). Long-term communities, which correspond to coarse levels of temporal resolution called biofacies ("community-type") by McKinney et al. (1996), and even coarser units called ecological-evolutionary units (EEUs) by Boucot (1983, 1990; figure 5.4), turn over only when significant disturbance occurs (such as extinction; see later in this chapter and in chapter 8). These communities are the ones that have typically been studied by paleontologists. The biofacies and EEUs are sufficiently coarse to damp any "noise" produced by the turnover of rare species (McKinney et al., 1996).

The argument about stasis in the fossil record is related to a longstanding debate about whether biological communities consist of highly interdependent species or a haphazard sample of species that just happen to occupy the same environment at the same time (Jack-

FIGURE 5.4 Ecological and evolutionary units (EEUs) of Boucot (1983, 1990) through the Phanerozoic (I–XII) for benthic marine communities in general. Each lettered interval (A–G) represents a time of reefs with complex frameworks. Hachures in lettered intervals indicate lack of reef frameworks, whereas hachures under diversity curve for well-skeletonized taxa indicate reorganization of Boucot's EEUs (I–XII). Arrows denote mass extinctions (see chapter 8). Diversity of evolutionary faunas (1–3) of Sepkoski and Miller (1985) are also indicated (cf. figure 5.1). Note that both reef and soft-bottom communities tend to turn over simultaneously. *From P. Sheehan, 1985. Reefs are not so different: they follow the evolutionary pattern of level-bottom communities. Geology 13:46–49. Reprinted with permission of the Geological Society of America.*

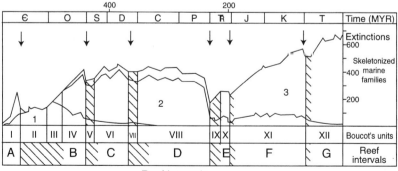

Reef Intervals

son, 1994, and references therein). Some workers have suggested that the latter view is simply a result of observations of local colonization and extinction made over short ecological timescales, and that longer timescales of observation (such as those available from the fossil record) demonstrate otherwise (Bretsky and Klofak, 1986).

Buzas and Culver (1994) tested these hypotheses using the stratigraphic record of benthic species of foraminifera in Cenozoic deposits of the Atlantic Coastal Plain formed during different highstands of sea level. According to their reasoning, whenever sea level transgresses onto the continental shelf, new habitats are opened to invading species. Presumably these habitats are roughly equivalent for each transgression. If species composition of communities is fairly stable, then basically the same species groupings should occur during each highstand. What they found was that the species composition of foraminiferal "communities" varied significantly through time because many species either emigrated (went "extinct" locally) or immigrated from elsewhere. Some emigrants returned during later transgressions, but many did not.

The same dynamic relationship is hinted at by the high species diversity of foraminifera in tropical carbonate environments. Up to 150 species have been described from coral reef and related habitats of the Caribbean, many of which occur in the sediment of the same locale (see Cushman, 1921; Martin and Liddell, 1988; Martin and Wright, 1988; Culver, 1990). Certainly, some are dominant and persist (such as *Archais* and *Amphistegina* in back reef lagoons and fore reef environments, respectively), but others are rare. What could all those species possibly be doing there at the same time, unless the niche space among species is highly subdivided (dependent, say, on critical levels of different micronutrients)? Even then, during any one sampling period not all species are detected.

On the other hand, Pandolfi (1996) has found that the species composition of coral reef communities remained fairly constant during repeated sea-level changes of the latest Pleistocene (30–125 ka). Although reef ecosystems are dynamic over short temporal and spatial scales (chapter 1), Pandolfi's findings suggest that reefs may be fundamentally different from other ecosystems or that "different ecological patterns and processes occur at different temporal scales" (Pandolfi, 1996:152). Perhaps ecological dynamics also vary according to taxa, as suggested by the different results for corals and foraminifera both from tropical carbonate environments. Pandolfi's results are based on presence–absence data, however, and as we have seen in the particle

113

model, if a species is present in a fossil assemblage, it is more likely to be common, whereas a rare species is more likely to be absent from a fossil assemblage.

DISTURBANCE AND SUCCESSION

Despite apparent stasis in much of the fossil record, ecosystems are constantly recovering from disturbance (Reice, 1994). Virtually everyone has observed ecological succession, whether they know it or not. If a field is left fallow for a few years, grasses and small shrubs begin to appear. If this "pioneer" stage is allowed to continue further, small and eventually large species of trees emerge, so that after about a century or so, in temperate climes at least, a final climax community is reached, in which, apparently, the process of succession has stopped. As succession occurs, then, the organisms are conditioning and changing the environment. According to Odum (1969), these changes include high total community photosynthesis (gross production) and net photosynthesis (total photosynthesis minus total respiration), increasing biomass (total living tissue) and standing crop (number of organisms or population size), food chains becoming more complex (from linear to weblike), increasing diversity of animals as spatial heterogeneity (new habitat) develops, greater niche specialization (less overlap between the roles of different species in a community), and increased nutrient recycling within the community (table 5.2). If a disturbance occurs, say a forest fire caused by lightning, the forest will be set back to an earlier sere, or stage and is called a disclimax, from which it develops toward the climax stage once again. It is not my intent to discuss whether Odum's (1969) view of ecologic succession is the right one (see also Connell and Slatyer, 1977; Karlson and Hurd, 1993). Rather, it is whether these sorts of changes may be extrapolated to much longer timescales.

A number of workers have attempted to do so. Nicol (1962), for example, described "succession" in Silurian (Niagaran series) reefs of the midwestern United States. Other workers have continued in this vein for both reefs and soft-bottom communities (e.g., Walker and Alberstadt, 1975; see Dodd and Stanton, 1981, for further references). Copper (1988), for example, concluded that ecologic succession in individual Phanerozoic reefs took about 3000 years, much longer than for a temperate forest ecosystem (up to about one century). He also described what he called erathemic succession, which, although it follows the general pattern of ecologic succession in reefs, represents the global advance and retreat of reef ecosystems *as a whole* to global climate change over intervals of 30 to 100 Ma. By contrast, Miller (1986) suggested that reef development consists of a hierarchy of developmental processes, among which is ecologic succession (table 5.1).

Although fossil assemblages may record many spatio-temporal scales of development (from ecologic interactions upward), the processes of ecologic succession are not necessarily equivalent to those of community evolution, which occurs over much longer timescales (see Miller, 1986). For example, approximately three decades ago Helen Tappan (1968, 1971, 1982, 1986) began a series of classic papers on the interaction of the marine and terrestrial realms and their influence on nutrient input to the oceans, marine productivity, and plankton evolution and extinction. Basing her conclusions on a strict analogy to the phenomena of ecologic succession proposed by Odum (1969; table 5.2), Tappan (1971) described a sequence of Phanerozoic communities that was at odds with the fossil record. According to her (1971), youthful marine ecosystems are characterized by high primary productivity, low nutrient conservation (low rates of nutrient recycling), low diversity, an emphasis on "r-strategies" (opportunism), and short food chains. Mature ecosystems, on the other hand, are characterized by low primary productivity and high nutrient conservation, high species diversity, an emphasis on "k-strategies" (slow growth, delayed reproduction), and long food chains. Most importantly, according to Tappan (1971), ecosystems matured during the Paleozoic: Food chains became more weblike and the biomass (carrying capacity) of the global marine ecosystem increased (productivity/biomass ratio decreased).

Martin (1995a, 1996a, 1996b) re-examined Tappan's conclusions in light of later microfossil, stable isotope, and lithologic data (see chapter 7). Although he agreed with Tappan's inferences regarding secular increases in diversity, biomass, and the weblike nature of food chains, he disagreed with the results of her extrapolation of ecological succession to the Phanerozoic. He argued that in the post-Devonian Paleozoic, long food chains with well-developed predators (characteristic of mature ecosystems) were sustained by elevated nutrient recycling (characteristic of mature ecosystems) and *high* marine productivity (characteristic of youthful ecosystems). The Permo-Carboniferous was characterized by glaciation, enhanced ocean overturn, sea level fall, and extensive soil formation, all of which would have heightened nutrient input, primary productivity, and biomass in the photic zone (see also Pitrat, 1970). In contrast, Cambro-Devonian seas were largely

TABLE 5.2 Attributes of Ecological Succession

Ecosystem Attributes	Developmental Stages	Mature Stages
Community Energetics		
Gross production community respiration (P/R ratio)	Greater or less than 1	Approaches 1
Gross production standing crop biomass (P/B ratio)	High	Low
Biomass supported/unit energy flow (B/E ratio)	Low	High
Net community production (yield)	High	Low
Food chains	Linear, predominantly grazing	Weblike, predominantly detritus
Community Structure		
Total organic matter	Small	Large
Inorganic nutrients	Extrabiotic	Intrabiotic
Species diversity–variety component	Low	High
Species diversity–equitability component	Low	High
Biochemical diversity	Low	High
Stratification and spatial heterogeneity (pattern diversity)	Poorly organized	Well-organized

	Life History	
Niche specialization	Broad	Narrow
Size of organism	Small	Large
Life cycles	Short, simple	Long, complex

	Nutrient Cycling	
Mineral cycles	Open	Closed
Nutrient exchange rate between organisms and environment	Rapid	Slow
Role of detritus in nutrient regeneration	Unimportant	Important

	Selection Pressure	
Growth form	For rapid growth (r-selection)	Longer-lived (K-selection)
Production	Quantity	Quality

	Overall Homeostasis	
Internal symbiosis	Undeveloped	Developed
Nutrient conservation	Poor	Good
Stability (resistance to external perturbations)	Poor	Good
Entropy	High	Low
Information	Low	High

Modified from Odum, 1969.

characterized by low nutrient recycling (youthful ecosystems) and *low* productivity (mature ecosystems), which were able to sustain only short food chains (youthful ecosystems). We return to this topic in chapter 7, where the relationship between energy (food) and evolution is examined.

ALTERNATIVE COMMUNITY STATES

Based on the fossil record, the recovery of marine communities following a mass extinction has largely been one of assembling newly evolved taxa (descendants of the survivors) into new ecosystems (chapter 8). Reefs are recognizable as reefs, for example, but stromatoporoids (which dominated early-middle Paleozoic reefs) were eventually supplanted by scleractinian corals. Given evolutionary constraints (the fact that it takes a long time for new species to evolve), however, communities cannot respond in a similar manner over much shorter (ecological) timescales. Nevertheless, they may substantially change their character in response to natural or anthropogenic change.

Based on the model of May (1977), Knowlton (1992) concluded that marine ecosystems, especially reefs, are *metastable;* they exhibit different community states under a single physical environmental regime (see also Knowlton et al., 1990; Done, 1992). Alternative states are usually the result of some disturbance or perturbation. Such behavior has very important implications for humankind's ability to predict the response of reefs and other ecosystems to natural versus anthropogenic disturbance and to conservation (management) efforts.

These multiple stable community states are known as attractors in chaos theory (figure 5.5), the "study of unstable aperiodic behavior in deterministic nonlinear dynamical systems" (Kellert, 1993:2). Most people associate chaos with randomness, but the term *chaos* actually refers to unpredictable systems that are *deterministic,* in which "future states are completely fixed by [a system's] current state and its rules of dynamical motion" (Casti, 1994:87; see chapter 3 of Casti, 1994, from which much of the following discussion comes; Kellert, 1993, gives a more technical but mostly nonmathematical account). One of the best examples of a deterministic dynamic system (a system that can be described and predicted mathematically) is the movement of the planets within the solar system, in which current planetary positions can be used to calculate future (and past) positions using Newton's laws of motion. (How else would astronomers be able to predict eclipses and other astronomical phenomena? Because ancient civilizations also performed similar feats, *without* Newton's laws, these phenomena must indeed be highly deterministic.) But *deterministic* does not necessarily mean predictable. Chaotic systems are deterministic, but they behave unpredictably, which is why they appear to behave randomly. However, truly random systems are not deterministic.

There are several types of attractors. In the simplest case, the system "spirals" to a steady-state, as a ball in a funnel does, for the given set of conditions (parameters) and is called a fixed-point attractor (figure 5.5a). A fixed-point attractor is said to be aperiodic or constant because if such a system is disturbed, it spirals back to the fixed point as long as the system remains within the domain of attraction of the fixed point.

In the simplest type of limit cycle, the system moves periodically between two different states, like a ball in a sombrero or an ashtray (figure 5.5b). One of the best examples is the fluctuation in numbers of a population, which can be described by the logistic equation:

$$dN/dt = rN(1 - N/K), \qquad (5.4)$$

where N = the number of organisms, r = the intrinsic rate of population increase, and K = carrying capacity of the environment (the maximum number of organisms that the environment can sustain; Casti, 1994, gives other examples of phenomena that can be described by the logistic equation). The term $1 - N$ acts as a negative feedback on birth rates that damps the system as a result of overcrowding and competition for increasingly scarce resources.

The behavior of a population can vary drastically depending on how the parameters of the equation are varied. The term *parameter* refers to a term in a function that determines the *form* of a function, but not its general nature. In the equation $y = mx + b$, for example, m and b are parameters that determine the slope and y-intercept of a straight line, whereas x and y are variables that are linearly related (that is, a change in x produces a proportional change in y). The equation always describes a line, but depending on the parameters, the slope and y-intercept vary. Changing the parameters of an equation or system is kind of like fiddling with the knobs on a TV set, sometimes with drastic results, especially in chaotic systems. Thus more complex limit cycles of period 4 or more may occur (figure 5.6c).

Nevertheless, fixed-point attractors and limit cycles are normally locally stable; that is, small changes in the parameters or conditions make only minor changes in the end result (think about varying the

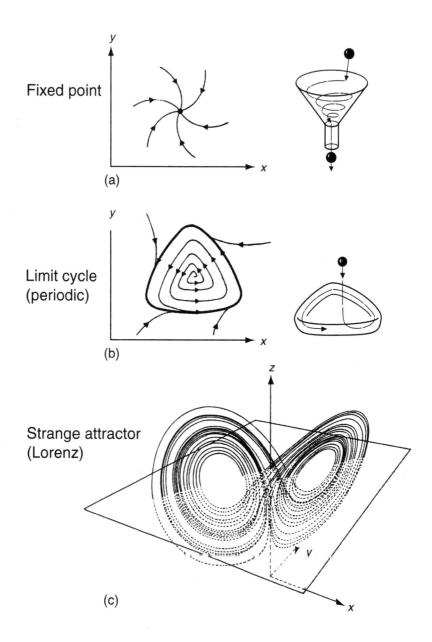

Fixed point

(a)

Limit cycle
(periodic)

(b)

Strange attractor
(Lorenz)

(c)

FIGURE 5.5 Types of attractors. (a) In the case of a fixed-point attractor, if
a system is disturbed, but remains within the indicated domain of attraction, it will eventually move (spiral) back to its original state (point), much

like a ball in a funnel, because the system is at equilibrium (steady-state) for the given set of conditions. (b) In limit cycles, the system moves periodically between two or more different states, like a ball in an ashtray. One of the best examples is the fluctuation in numbers of a population. Note that whether the system starts inside or outside the domain of attraction of the limit cycle, the system moves to the attractor. (c) Strange attractors are actually the most common type of attractor, and consist of a combination of aperiodic and periodic pathways rolled into one. Unlike fixed points and limit cycles, the pathways along the strands are unstable, meaning that they repel one another rather than attract. Even a slight shift from one pathway to an immediately adjacent one causes a significant change in the trajectory of the system (Casti, 1994). This particular type of strange attractor is called a Lorenz attractor, the shape of which gave rise to the term *butterfly effect* (see text for further discussion). *Modified from Coveney and Highfield, 1990.*

FIGURE 5.6 Behavior of the logistic equation. (a) Steady-state (fixed point). (b) Two-cycle behavior. (c) Four-cycle behavior. (d) Chaos. *Modified from Coveney and Highfield, 1990.*

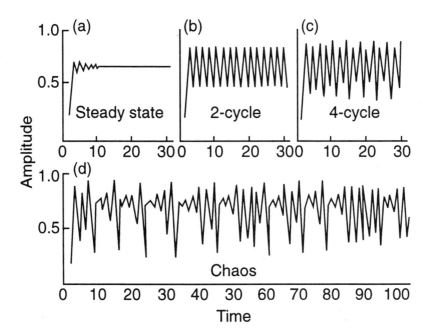

ingredients only slightly in a cooking recipe; most people will not notice the difference unless an ingredient is altered radically). The system typically returns to its state before the disturbance. Under appropriate conditions, however, chaos may also ensue in the logistic system (figure 5.6d).

Depending on the parameters, then, apparently random behavior ensues from a very simple system. Although the behavior is unpredictable, it is not truly random; it is actually highly deterministic. Chaotic behavior ultimately results because the system contains nonlinear (nonproportional) terms, as opposed to linear ones. Note, for example, that in equation 5.4 the parameters r and N are multiplied by themselves. This is what introduces the potential nonlinear behavior of the system. Other examples of nonlinear terms include sine (x) (Kellert, 1993).

Chaotic behavior is especially well-represented by the third type of attractor, the strange attractor (figure 5.5c). This is actually the most common type of attractor, and consists of a combination of aperiodic and periodic pathways rolled into one. Think of a ball of taffy (with raisins embedded in it) that is being pulled and stretched, then folded back on itself, only to be pulled and stretched, folded again, and so on. Each layer of taffy may be thought of as the pathway of a dynamic system. As the taffy is being manipulated, raisins that initially started very close together in the original taffy ball may suddenly move along pathways that rapidly become very far apart and vice versa. This is precisely the behavior on a strange attractor. Unlike those of fixed points and limit cycles, the pathways of the strange attractor are unstable, meaning that they repel one another rather than attract. Even a slight shift from one pathway (taffy layer) to an immediately adjacent one will cause a significant change in the trajectory of the system (Casti, 1994). Unlike fixed points and limit cycles, strange attractors are fractal (self-similar; remember that the pathways were folded back on one another in the taffy example) and display sensitive dependence on initial conditions; that is, the behavior of the system depends on where it started (Coveney and Highfield, 1990), and chaotic systems with only infinitesimally small initial differences may evolve radically different behaviors. This behavior is also known as the butterfly effect (named by meteorologist Edward Lorenz in the early 1960s, who used dynamic systems on digital computers to model weather—no small feat considering the state of computers at the time). According to this effect, the flapping of a butterfly's wings in Brazil may make the difference between calm and stormy weather in Texas days or weeks later

(Kellert, 1993), which is why the weather is so unpredictable more than a few days in the future.

If disturbance is normal and communities are constantly recovering from some sort of ecologic disturbance (Reice, 1994), then, given the widespread and often subtle impact of humans on the environment (see Runnels, 1995), how are we to determine what is natural about a community and what is not? The answer is, of course, the fossil record because it represents the long-term response of biota to environmental change before human interference. The more often a particular community state occurs in the fossil record (the larger its domain of attraction), the more often it should occur in the fossil record. Based on the record of Pleistocene reefs in the Caribbean, for example, Jackson (1992) found that reefs were dominated by the coral genera *Acropora* and *Montastrea,* both of which are dominant in undisturbed (normal) reefs today (but refer back to the section on stasis and community unity).

These attractors (community states), then, are typically stable and the fossil record seems to indicate that a massive disturbance is needed to shift ecosystems to a new community type or attractor (again, see the section on stasis and community unity). Nevertheless, several minor disturbances to the reefs at Discovery Bay, Jamaica, beginning with Hurricane Allen in 1980 and followed by a massive die-off of sea urchins that feed on algae, have significantly altered the reef community there. Because macroalgae grow faster than corals (and their spat, or settled larvae), loss of the urchins has allowed algae to overwhelm the reefs. As of 1995, the Discovery Bay reefs are dominated by algae, and it is uncertain whether the reefs will return to their previous (and presumably normal) coral-dominated state. Apparently the disturbances were of sufficient magnitude to shift the community to an alternative community state, at least for the foreseeable future. In a similar example, reefs on the north coast of Cuba were smothered with carbonate sands as part of a beach nourishment project. Over the succeeding 20 years, however, the sands have been eroded away and hard substrates, which are preferred by corals, are becoming increasingly available. Consequently, the reefs are gradually restoring themselves, presumably through settlement of larvae from healthy populations located further offshore (Iturralde-Vinent, 1995; personal observations by the author in 1995). Unfortunately, these and other reefs in the vicinity may soon be threatened by hotel and condominium development and the accompanying sewage and sedimentation.

Indeed, rising nutrient levels in many localities may be threatening the health of coral reefs or have already largely destroyed them (Smith et al., 1981; Tomascik and Sander, 1985, 1987a, 1987b). Increased nutrient levels stimulate plankton production and the invasion of rapidly growing suspension-feeding organisms that outcompete corals and shift communities toward ones dominated by macroalgae and sponges (Hallock and Schlager, 1986; Hallock, 1987; Hallock et al., 1993, 1995). Although Szmant and Forrester (1996) recently concluded that nutrient levels are not elevated in waters off the Florida Keys, their samples were taken during a 1- to 2-year period. By contrast, working in Biscayne Bay (Florida), Hudson et al. (1994) found that growth rates calculated from colonies of *Montastrea annularis* (a common reef-building coral) as old as 200 years began to decline after 1950 and that the decline may be related to increasing sewage effluent; Bell (1991, 1992) and Swart et al. (1996) came to similar conclusions for the Great Barrier Reef (Australia) and Florida Bay, respectively. Note how the contrasting conclusions about the role of nutrients and reef health depend on the scale of temporal measurement (chapter 1); there is much food for thought here for all environmental studies.

We return to upward scaling of ecological processes, attractors, and a related phenomenon called bifurcation in chapters 7 and 8, and examine the effects of environmental disturbance on life-history traits, geographic distribution, and speciation and extinction of corals.

THE ORIGINS OF PHYLA

So far we have discussed a variety of biological phenomena, all the while bearing in mind conflicting interpretations that arise from considering different spatio-temporal scales.

What about the origins of higher taxonomic categories? For example, why did no new major invertebrate phyla, which are characterized by distinctive body or "ground" plans (all molluscs have the same basic features, for example), evolve after the Cambrian? Gould (1989) argued that during the Cambrian radiation, when abundant fossils began to appear, creatures (such as the bizarre ones of the Burgess Shale) experimented with all sorts of ways to fill the niches and habitats that lay wide open to them at this time in Earth's history. Life was radiating into a frontier, like the settlers in covered wagons racing to stake out claims in the Oklahoma land rush. Many of these new taxa, Gould maintains, cannot be shoehorned into the phyla recognized by systematists today and so they represent altogether different phyla with highly unusual (by our standards) body plans. Most of these experiments (phyla) failed, leaving behind the phyla that have persisted up to the present. As a result, the evolutionary tree of these phyla does not resemble a tree so much as a bush, with many branches at the base that were progressively pruned after the taxa had originated (see Gould, 1989, his figure 1.17; see also the discussion of species sorting earlier in this chapter).

This situation is analogous to the development of the bicycle. Early bicycles came in a variety of shapes and sizes, including the wheels and handlebars, but over the years several types have come to dominate: street, racing, and mountain bikes. Kauffman (1995) argues that this pattern is so widespread—from nature to the development of technological advances such as the computer—that it may actually represent "lawlike" behavior.

How could this have happened? Perhaps according to the genomic hypothesis. Mutations that normally occur early in the embryonic development of an organism are almost always detrimental because the effects cascade through the rest of ontogenetic (the individual's) development, which is a tightly controlled process. Early development tends to lock in later developmental pathways (such as the differentiation of new tissues and organs and their positions in the embryo) during individual development or ontogeny (Kauffman, 1993, 1995); such a system is thus hierarchical. But in the Cambrian, the organisms had not yet adapted to the ecological landscape, so ecological constraints were largely missing; the ecological and evolutionary landscape available to newly evolving species was wide open and virtually anything was possible, at least for a time. Surviving phyla were adapted to the environment, thereby providing constraints on the subsequent radiation of new taxa.

There is evidence that these sorts of constraints do indeed occur. In the 1960s François Jacob and Jacques Monod both received the Nobel Prize for their description of the operon model of protein induction for the enzyme β-galactosidase. Production of the enzyme is caused by the introduction of the sugar lactose to cultures of the gut bacterium *Escherichia coli*. Within a few minutes of addition of lactose, the bacteria begin to synthesize β-galactosidase and use lactose as a carbon source. Lactose acts as an inducer that turns on the gene that codes for the enzyme. Another protein is coded by another regulator site that acts as a repressor: It binds to the DNA at the operator site and blocks (turns off) synthesis of the messenger RNA transcribed from DNA and that would be translated into β-galactosidase via protein synthesis. Lactose binds to the repressor protein and alters its shape (which is very specific) and prevents it from binding to the DNA. Thus the gene for β-galactosidase is read and lactose is cleaved (see Stryer, 1975, figures 27-3 and 27-4).

Getting back to development, if changes were to occur to a master regulator or operator sites involved in development (known as homeobox or Hox/HOM genes) that controlled early aspects of ontogeny, such as which regions or groups of cells develop into certain organs or limbs (Davidson et al., 1995b), the whole system would presumably be disrupted. This would normally prove lethal, except, perhaps, if the organism had no ecological constraints to contend with, as in the Cambrian. On the other hand, according to this hypothesis, despite the great ecological vacuum left by the massive Late Permian extinctions (chapter 8), these developmental constraints would have been sufficient to constrain the recovery of new taxa; the survivors simply had too much genetic baggage to carry around from their ancestors to develop wholly new groundplans (Kauffman, 1995).

Despite the appeal of the genomic hypothesis, views of the Burgess Shale creatures have moderated somewhat because with further study the taxonomic relationships of some have been clarified and they are no longer seen as all being wholly unique phyla. Valentine (1995), for example, suggested that vacant ecospace following the end-Permian extinctions (chapter 8) was rapidly filled by lineages derived from pre-existing taxa so that no new distinctive groundplans evolved (ecospace hypothesis). The ecospace after the end of the Permian was not as wide open as the massive extinctions (and the fossil record) would have us believe, and the evolution of new taxa was constrained more by ecology than by development (Kerr, 1994). Moreover, Valentine (1995) maintains that genetic controls on development are not as inflexible as they might appear. He maintains that genetic changes in such constraints were crucial to the explosion of Cambrian taxa because the earliest forms all evolved from an ancestor with a single Hox/HOM cluster (see Erwin et al., 1997, for recent review).

TIME, LIFE, AND HUMANKIND: APPLICATIONS

So it appears that paleontology and stratigraphy are important to understanding concepts such as the assembly, structure, and persistence of biological communities. But all good capitalists—including university administrators—know that you cannot make money off paleontology (except if one still has a job with an oil company), so we now descend from the ivory tower and apply some of this knowledge through a brief overview of some examples of applied micropaleontology. The reader is encouraged to delve into the papers cited in this section for further examples.

Natural Resources: Oil and Water

The most obvious application of paleontology has been in oil and gas exploration, which has a long history beginning about 1915–1920 with the establishment of micropaleontological laboratories in oil companies in the United States. In a landmark paper published in 1925, micropaleontologists Esther Applin of Rio Bravo, Alva Ellisor of Humble Oil and Refining Company (now Exxon), who later became chief paleontologist for Humble, and Hedwig Kniker with the Texas Company (Texaco), revolutionized industrial micropaleontology by using foraminifera in subsurface correlation in Texas and Louisiana (Applin et al., 1925; see also Jones, 1956; Stuckey, 1978; and Martin and Liddell, 1991). Standardized zonations soon followed, and the relationship between micropaleontology, stratigraphy, and petroleum exploration was firmly established for the next half century or so.

Most recently, Armentrout (1991, 1996), among others, has used paleontology extensively to develop predictive models of reservoir sand occurrence that can save time, energy, and money in petroleum exploration. Despite the extensive use of seismic stratigraphy in petroleum exploration, it is still a crude tool in terms of refined stratigraphic zonation and correlation. Indeed, practical resolution on seismic sections is typically no better than 10–50 m (Hallam, 1992). Without paleontology, correlating seismic lines, especially when the data are being brought into a "wildcat" region (where no wells have previously been drilled) from another area far away, is tenuous, to put it politely. I once spent five days on a wildcat well examining well samples for a particular biostratigraphic marker because the geologist and geophysicist incorrectly correlated seismic lines into the wildcat region on the basis of the character of a particular seismic reflector. The well was also located over a geopressurized zone (caused by the great thickness, 3000–5000 m or more, of overlying sediment), which, if drilled, could have caused a blowout and loss of the drilling rig and life. Denne (1994) describes the use of graphic correlation to determine the approximate depths of stratigraphic markers as the well is being drilled and the proximity of geopressurized zones.

Ironically, despite the extensive use of micropaleontology in petroleum exploration, foraminifera were used much earlier to date strata in a water well in 1877 near Vienna, Austria. Further studies followed by other workers in the United States (see Jones, 1956:1–6, for references), but it was primarily J. A. Udden of Augustana College (Illinois) who, in 1911, began to stress the importance of microfossils in correlating water wells. Udden would later forsake academics to become head of the newly organized Bureau of Economic Geology of Texas, where he shifted application of microfossils from water to petroleum, and the discipline of micropaleontology began to flourish.

Since Udden's time, we seem to have come full circle (Martin, 1991, 1995b). Although most Americans take clean water for granted, it is just as precious as oil (try drinking the tap water in some countries). What is the stratigraphic occurrence of aquifers in a given area? Are they the same as the ones that occur in another section far away? The answers to these questions are needed if the aquifers are to be mapped, just as they are needed in petroleum exploration in mapping the distribution and thickness of often thin, highly discontinuous reservoir sands originally deposited in submarine fans. Sloan (1995), in an issue of the *Journal of Foraminiferal Research* devoted entirely to using foraminifera to solve environmental problems, describes the use of foraminifera (and other microfossils) to correlate discontinuous water-bearing units of alluvial origin in the San Francisco Bay area that serve as aquifers for municipal and agricultural water supplies.

Anthropogenic Effects on Ecosystems

Culver and Buzas (1995) review the use of foraminifera in studying the effects of human impact on ecosystems. They emphasize that foraminifera respond to such disturbances as global warming, as do other biota, but they emphasize that human disturbance of coastal habitats has also had quite a deleterious effect. Foraminifera are ideally suited to such coastal environmental studies because their short generation times allow them to respond quickly to environmental change (Alve, 1995). For example, Bandy et al. (1964a, 1964b, 1965a, 1965b) were among a handful of workers in the late 1950s and early 1960s who examined the response of foraminiferal populations to large inputs of sewage in shallow marine waters. Unfortunately, Bandy et al.'s studies lay dormant until fairly recently, probably because of the heavy emphasis of applied micropaleontology on biostratigraphy in petroleum exploration.

Also, foraminifera appear to have fairly specific cellular defense mechanisms against toxic substances (Bresler and Yanko, 1995) that, with further research, might prove useful in assessing pollution sources. Alve (1995; special issue of *Journal of Foraminiferal Research*) reviewed the recent use of foraminifera in studies of estuarine pollution caused by sewage, petroleum, heavy metals, and thermal effluent from power plants (other studies of the response of foraminifera to pollution are also presented in the same journal issue). Some opportunistic (or r-selected) species of foraminifera (those with high intrinsic rates of reproduction; see equation 5.4) are quite tolerant of abnormal conditions and so may bloom under stressed conditions. Indeed, Hallock et al. (1995) used changes in foraminiferal assemblages to detect long-term eutrophication in waters off Key Largo, Florida, as did Barmawidjaja et al. (1995) for the last 150 years in the Adriatic Sea. Tests (shells) of foraminifera may also be deformed in some cases as a result of pollution. Alve (1995) emphasized that dated cores can also be used to assess preanthropogenic conditions, but taphonomic processes such as dissolution must be considered in assessing preanthropogenic habitats.

Cooper (1995) used diatoms (siliceous microfossils) and pollen to reconstruct a 2000-yr history of Chesapeake Bay. She found that sedimentation, eutrophication, and anoxia (lack of oxygen related to massive diatom blooms caused by eutrophication and then die-off) all increased dramatically following settlement of the region. Increased sedimentation and eutrophication were related to land use (deforestation), sewage input, and freshwater runoff. It appears that Chesapeake Bay has shifted to a new stable state as a result of human activity.

Similarly, Smith (1995) has used lacustrine (lake-dwelling) ostracodes (microscopic crustaceans) to reconstruct air temperature and precipitation (regional hydrology) patterns. Among her contributions, she has demonstrated how regional drought patterns set in at the close of the Little Ice Age in the Great Plains about the mid-nineteenth century, how groundwater recharge systems collapsed during droughts in Nevada, and the distribution of pollutants in wetlands.

Numerous workers have used foraminifera or other microfossils to infer rates of sea-level change (see Scott and Medioli, 1980, 1986; John, 1995). Certain species are indicative of elevation above mean sea level and can be used to determine ancient positions of sea level in cores.

Recently, other workers have emphasized the effects of taphonomy and infaunal habit on the occurrence and interpretation of foraminiferal assemblages in sea level studies (Goldstein and Frey, 1986; Patterson et al., 1994; Goldstein and Harben, 1993; Goldstein et al., 1995). Hoge (1994) emphasized the role of biogeochemical cycling in the preservation of marsh foraminifera.

Hippensteel and Martin (1998) used vibracore records of storm-deposited benthic foraminifera to infer that a major source of sediment for barrier island construction off Charleston, South Carolina, was the shallow shelf seaward of the islands. Oligo-Miocene foraminifera were eroded from offshore outcrops by storms and deposited with sands in back-barrier marshes. They calculated that the storms were of gale size or larger; such data may be of interest to insurance companies. They also inferred rates of bioturbation by calculating slopes of upwardly smeared (bioturbated) storm layers (see chapter 3).

Bell (1992) suggested that rising nutrient levels may have contributed to outbreaks of the crown-of-thorns starfish on the Great Barrier Reef. The starfish then attack the corals and decimate the reefs. When the starfish die, they contribute significant numbers of skeletal elements to the sediment. One way to test for the occurrence of similar outbreaks in the past (before anthropogenic nutrient input) is to examine the sedimentary record, the results of which have been disputed (see Greenstein et al., 1995, for review). In order to clarify the situation, Greenstein et al. (1995) conducted field experiments that simulated outbreaks and mortality of the starfish. They found that after 4 years, the signature of the simulated outbreaks was best recorded by starfish ossicles of 1–4 mm in size. Numbers of ossicles 0.5–1-mm and above 4 mm in size increased and decreased, respectively, because of taphonomic bias (based on tumbling experiments) that influences any estimates of past populations sizes. Clearly, taphonomy must play a significant role in environmental management decisions based on the sedimentary record (see also chapters 2, 3, and 4).

Seismic Hazards and Engineering

Microfossils have also been used in engineering. Hart (1996) discussed the use of foraminiferal biostratigraphic zonations in aligning the Chunnel that now connects England and France beneath the English Channel as it was being excavated from both sides of the channel. Foraminifera have also been used to correlate alluvial deposits at seismic hazard sites in the San Francisco Bay area (Sloan, 1995).

Microfossil assemblages have even been used to study the frequency of earthquakes. Foraminifera, diatoms, and pollen have all been used to identify the original paleoenvironments (such as marine and freshwater peats, tsunami deposits, and soil horizons) of sediments uplifted along active margins. Coupled with radiocarbon dates, the frequency of large (magnitude 8–9) earthquakes may then be calculated over intervals of thousands of years (Nelson and Jennings, 1993; Mathewes and Clague, 1994; Hyndman, 1995).

6

Cycles and Secular Trends

There are those who try to generalize, synthesize, and build models,
and there are those who believe nothing and constantly call for more
data. The tension between these two groups is a healthy one; science
develops mainly because of the model builders, yet they need the
second group to keep them honest. —Andrew Miall (1990)

Seeing is believing: I wouldn't have seen it if I didn't believe it.
—Anonymous

The second quote pretty much sums up many geologists' attitudes
about the existence of cycles in the geologic record, but it is the first
quote from Miall that is most constructive. Many workers continue
to pursue evidence of cyclicity in the fossil record despite the skepti-
cism of their colleagues. If a phenomenon is cyclic, then we ought to
be able to predict (forecast) its next occurrence ("retrodict," or hind-
cast, if one is looking into the past). When most geologists think of
cycles, they automatically think of the rise and fall of sea level and its
relationship to other disciplines, such as petroleum exploration (chapter 2). Consequently, a vast and daunting literature has developed on
sea level.

But many other cycles and associated phenomena are apparent in
the geologic record. The cycling of elements in the earth's crust and
shallow mantle by the interaction of physical and biological entities
has played a critical role in controlling the earth's climate, and, as discussed in chapter 7, the evolution of the earth's biosphere. Nowhere
is the action of different processes on different timescales more apparent than in these biogeochemical cycles.

CYCLES OF SEA LEVEL

We consider cycles of global (eustatic) sea level first because they have often been viewed as a causal agent of change in the earth's history, although they are themselves often tied up with other cycles. Sea-level change consists of cycles superimposed on cycles, and we consider the broadest changes first. A supercontinent insulates the earth, slowing radiogenic heat loss from the core and mantle. Eventually heat buildup between continental crust becomes sufficient to cause doming, rifting, and breakup of the continent, followed by rapid seafloor spreading. Eventually the continents collide and become sutured together to form yet another supercontinent and then the process begins again. This cycle of supercontinent assembly–rifting–assembly has occurred at least four times in the last 2 billion years (Giga-annums, or Ga).

Fischer (1984) subdivided the Phanerozoic into two of these tectonically driven supercycles (each of about 300 Ma long) of continental configuration and sea level (table 6.1; figure 6.1; note that Fischer's use of the term *supercycle* differs from that of other workers mentioned in

TABLE 6.1 Stratigraphic (Sea Level) Cycles and Their Causes

Type	Other Terms	Duration (Ma)	Probable Cause
First order	—	200–400	Major eustatic cycles caused by formation and breakup of supercontinents
Second order	Supercycle, sequence	10–100	Eustatic cycles induced by volume changes in global midoceanic spreading ridge system
Third order	Mesothem, megacyclothem	1–10	Possibly produced by ridge changes and continental ice growth and decay
Fourth order	Cyclothem, major cycle	0.2–0.5	Milankovitch glacioeustatic cycles, astronomical forcing
Fifth order	Minor cycle	0.01–0.2	Milankovitch glacioeustatic cycles, astronomical forcing

Modified from Miall, 1990, after Vail et al., 1977.

133

table 6.1). In the Phanerozoic phases of these cycles, during much of the Cambrian-to-Devonian and Jurassic-to–Late Eocene, continents were dispersed and shallow, widespread epeiric seas rose onto continents as a result of increased seafloor spreading rates and mid-ocean ridge (MOR) volume. Conversely, sea level underwent a broad decline during the Late Devonian to Triassic and from the Oligocene to the Recent.

The Triassic to Recent is largely characterized by the rifting apart of Pangaea (Greek for "all land"). How could sea level fall from the Oligocene on if most of the continents are moving away from each other? Because superimposed on these broad tectonically driven changes in sea level are smaller cycles (or trends, in the case of the Oligocene-to-Recent fall, because this "cycle" has yet to be com-

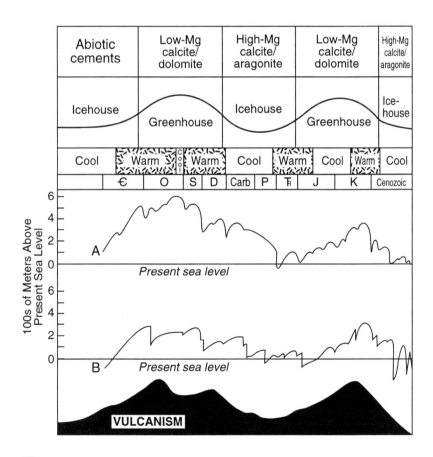

pleted). As Pangaea rifted apart, Antarctica and Australia moved away as a single continent. Ultimately, Antarctica and Australia separated and moved toward their present positions. By the Late Eocene, it appears that Antarctica was over the South Pole, because there was a major cooling and biotic turnover (as evidenced by benthic foraminifera) in the deep ocean or "psychrosphere" (the root *psychro* refers to cold). The waters adjacent to Antarctica apparently became sufficiently cool (and dense) to sink and set up a new pattern of deep-ocean circulation. Ice buildup on the continent may also have contributed because when ice forms from seawater, the salt is left behind, which also increases the density of the remaining water. This kind of ocean circulation is called thermohaline (temperature- and salinity-driven) circulation (discussed in more detail later in this chapter). Changes in MOR volume may also explain second-order changes in sea level of 10–100 million years' duration (table 6.1), which includes the cratonic sequences (such as Sauk and Tippecanoe) recognized by Sloss (1963) in North America and by workers on other continents.

The cause of third-order cycles is much more ambiguous. Changes in MOR volume are probably too slow to account for third-order cy-

FIGURE 6.1 Phanerozoic cycles of continental configuration, sea level, and climate. During greenhouses (Cambro-Devonian; Jurassic–Late Eocene), continents were dispersed and sea level raised as a result of increased seafloor spreading rates and mid-ocean ridge (MOR) volume. Increased volcanism (based on magma emplacement) resulted in increased atmospheric pCO_2 that presumably caused the earth's average surface temperature to rise during these intervals. Just the opposite conditions apparently prevailed during icehouses (Late Devonian–Triassic; Oligocene–Recent), when atmospheric pCO_2 and sea level declined, presumably as a result of decreased spreading rates and MOR volume. Deposition of low-Mg calcite (in cement and ooids) during greenhouses presumably reflects increased atmospheric pCO_2 and decreased surface ocean $CaCO_3$ saturation and accelerated diagenesis. During icehouses, precipitation of high-Mg calcite and aragonitic ooids and cements prevailed. Atmospheric pCO_2 was, overall, probably much lower after the Cambro-Devonian because of increased terrestrial photosynthesis. Shorter cool and warm episodes after Frakes et al. (1992). Sea level after (A) Hallam (1992) and (B) Vail et al. (1977). *Reprinted from* Global and Planetary Change, *volume 11, Martin, R. E., Cyclic and secular variation in microfossil biomineralization: Clues to the biogeochemical evolution of Phanerozoic oceans, pp. 1–23, 1995, with kind permission of Elsevier Science–NL, Sara Burgerhartstraat 25, 1055 KV Amsterdam, The Netherlands.*

clicity. Third-order cycles seem to occur on a regional basis, but can be correlated from one region to another, so that regional tectonic events caused by changes in seafloor spreading rates (regional plate rifting and convergence) and intraplate stresses (Cloetingh, 1988) may be involved (Miall, 1990; see also Hallam, 1992).

However, changes in seafloor spreading rates (a few centimeters per year on average, or about how much one's fingernails grow in a year) and MOR volume are far too slow to account for higher-frequency cycles. Cycles of solar insolation (and accompanying waxing and waning of glaciers) more likely drive sea-level fluctuations at the much shorter Milankovitch frequencies (table 6.1), calculated by Serbian mathematician Milutin Milankovitch by about 1920. These cycles are the result of minor changes in the gravitational attractions between the planets and are of three basic types at the level of fifth-order cycles. The first cycle (eccentricity or ellipticity) is concerned with the shape of the earth's orbit about the Sun, which changes from circular to very slightly elliptical and back again over the characteristic period of 100,000 years. Glaciation is more likely to occur when the earth's orbit is most elliptical. Longer, eccentricity-related variations of 400,000 and 2 million years have also been demonstrated (Kerr, 1991; Krumenaker, 1994), but their exact significance remains unclear.

The second period is of about 40,000 years duration and is driven by changes in the angle of the earth's axis of rotation (obliquity or tilt) of about 22–25° from the vertical (figure 6.2). The northern hemisphere of the earth receives the greatest amount of sunlight when the earth is tilted maximally toward the Sun, and the least when it is tilted away. All gradations occur between the angles of 22° and 25° from the vertical.

The third period of 19,000–23,000 years is caused by progressive shifts (precession, or wobble) in occurrence of the earth's seasons on its orbit around the Sun, and accounts for most of the change in incoming solar radiation through time (figure 6.2). Each year the northern hemisphere of the earth passes through a summer solstice at about June 21, when the maximum intensity of solar radiation is highest in the northern hemisphere and we experience the season of summer (a corresponding winter solstice occurs in the southern hemisphere). After June 21, the northern hemisphere begins to tilt away from the Sun; the change is *not* the result of a change in obliquity, which occurs over a much longer timescale, but the result of a shift of the earth's position in its orbit. By about September 22, the Sun's rays have, in effect, migrated southward over the earth's surface and become concentrated at the equator, hence the autumnal equinox, when the concentration of insolation is equal in both hemispheres. As the earth moves further along in its orbit, the net effect is to make the northern hemisphere tilt even farther from the Sun and we eventually experience winter solstice (or summer solstice in the southern hemisphere) about December 21. Then the earth moves gradually toward its position at about March 20 (spring equinox), by which time, of course, the northern hemisphere is warming up and the southern cooling down. Note that today, winter solstice, when the northern hemisphere is tilted away from the Sun, occurs when the earth's orbit is *closest* to the Sun (perihelion), and that summer solstice, when the earth is

FIGURE 6.2 Effect of tilt and precession of the earth's axis of rotation on the distribution of sunlight reaching the northern hemisphere. The degree of seasonality increases with increasing tilt; the current angle of tilt is 23.5°. The precession of the equinoxes (or wobble of the earth's axis of rotation) also causes the amount of solar radiation hitting the earth to vary. The positions of summer and winter solstices (June 21 and December 21) and autumnal and spring equinoxes (March 20 and September 22) shift around the earth's orbit and complete one cycle about every 22,000 years. *From T. H. van Andel,* New Views on an Old Planet. © *Cambridge University Press 1985, 1994. Reprinted with the permission of Cambridge University Press.*

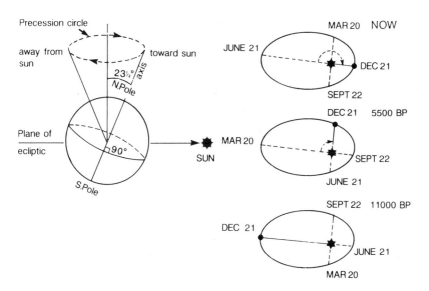

tilted toward the Sun, occurs when the earth is farthest away. This demonstrates the importance of the precession of the equinoxes and obliquity over eccentricity in determining overall solar input to the earth's surface.

This description of Milankovitch cycles applies only to today and the fairly recent past (last few thousand years), however. Before this time, the positions of the solstices and equinoxes differed significantly from those of today (figure 6.2). For example, 11,000 years ago, the summer solstice occurred slightly closer to the earth than it does now. Northern hemisphere ice caps began a general retreat about this time, paving the way for the spread of human culture and giving rise to our own anthropocentric view of the world encompassed by the term *Holocene.*

Although the cycles of insolation were calculated in theory by Milankovitch, they were eventually confirmed using the oxygen isotope record (figure 6.3). For our purposes, oxygen is represented by two stable (as opposed to radioactive) isotopes (elements that have the same number of protons, but different numbers of neutrons, and therefore different atomic weights). Oxygen-16 (^{16}O) has an atomic weight of 16 and is slightly lighter than its heavier counterpart, ^{18}O, which has two more neutrons. Although isotopes do not differ substantially in their chemical behavior, the weight difference between oxygen isotopes is enough to affect the formation of ice. ^{16}O is preferentially incorporated into ice as it forms, thereby enriching seawater in ^{18}O.

Organisms such as planktic foraminifera calcify as they live in seawater. In most instances the ratio of the two isotopes in fossil shells is quite similar or even identical to that in the seawater in which the animal lived (barring "vital effects" such as those produced by photosynthetic algae in some species, which can substantially alter the ratio):

$$Ca^{2+} + HCO_3^{-18/16} \rightarrow CaCO_3^{18/16}\downarrow + H^+. \qquad (6.1)$$

The ratio of oxygen isotopes (denoted $\delta^{18}O$), which is measured in concentrations of parts per thousand (‰), can be measured in a mass spectrometer from the CO_2 released by dissolving the shells in acid (the CO_2 got into the shells in the first place by dissolving into seawater to form carbonic acid, which then dissociates to form bicarbonate and carbonate ions):

FIGURE 6.3 Oxygen isotope curve for the last half of the Pleistocene Epoch plotted against core depth and magnetic polarity epochs. Odd numbers refer to warm (interglacial) episodes indicated by negative ("lighter") excursions in δ¹⁸O. *Modified from Imbrie and Imbrie, 1979, based on Shackleton and Opdyke, 1973.*

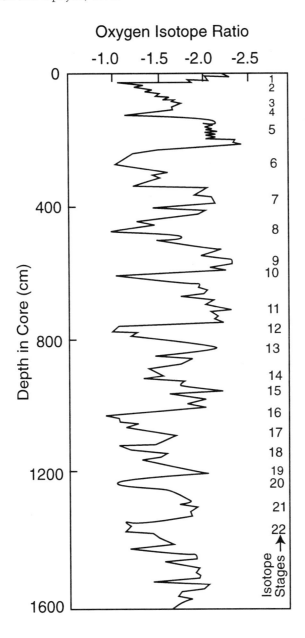

$$CO_2 + H_2O \rightarrow HCO_3^- \rightarrow H^+ + CO_3^{2-}. \qquad (6.2)$$

In determining the ratios using the CO_2 released in a mass spectrometer, then, we are basically reversing equation 6.1. These ratios may then be plotted against core depth or time (figure 6.3) if sediment accumulation rates are known with sufficient accuracy (chapter 2). The more negative (or "lighter") the ratio, the more ^{16}O was present in seawater and presumably the warmer the conditions. Based on the scheme devised by Cesare Emiliani, who was largely responsible for developing this technique under his major professor, Nobel Laureate Harold Urey at the University of Chicago, warm intervals are denoted by odd-numbered stages and cool ones by even numbers. This scheme works well until about the middle of the Pleistocene epoch, approximately 800,000 years ago, when the character of the isotope curve changes because of a shift from obliquity-dominated to eccentricity-dominated Milankovitch cycles, possibly as a result of chaotic behavior (van Andel, 1994; Krumenaker, 1994). Note the change in character of the oxygen isotope stages beginning about the middle of the Pleistocene (figure 6.3; see also figure 6.4); stages after about 0.7 Ma (stage 22) are much more distinctive than those before and therefore have been widely used in chronostratigraphic frameworks (chapter 2).

Note that the isotope curve of figure 6.3 is basically a summation of the three Milankovitch frequencies (figure 6.4; compare the character of the isotope curve to that of figure 5.6d). Two extreme examples are shown in figure 6.5. Indeed, it is these three frequencies that are found to be statistically significant in spectral analysis of the isotope record. Spectral analysis is a technique that recognizes statistically significant frequencies (such as those of climatic cycles) in time series data, such as the oxygen isotope record (sediment thickness is converted to time using calculations of sediment accumulation rate).

The recognition of sea-level cycles, especially the Vail curve (see figure 6.1), has been extremely controversial. First, the mechanisms causing sea-level change are not well understood. Second, most of the sections originally used by Vail et al. (1977) appear to have been from the Atlantic–North Sea regions and on craton margins or divergent, trailing-edge margins. Third, unconformities originally recognized on seismic sections were the basis for the recognition of sea-level curves (actually coastal onlap curves), even though unconformities are not indicative of the magnitude of sea-level change. Unconformities are obviously diachronous, yet they were correlated between regions. Miall (1990) and Hallam (1992) review other criticisms of the approach.

140

Nevertheless, seismic stratigraphy, with its attendant interpretations regarding eustatic sea-level change, and its offspring, sequence stratigraphy, which is less closely tied to the connotations of seismic stratigraphy, have become standard tools in petroleum exploration in developing predictive, high-resolution stratigraphic models of reservoir sand occurrence (see Armentrout, 1991, 1996; Martin et al., 1993; Martin and Fletcher, 1995).

Other workers have also used the sequence stratigraphic approach to develop high-resolution stratigraphic frameworks (see also chapter 2). Busch and West (1987), for example, used a hierarchical genetic stratigraphic framework consisting of third-, fourth-, and fifth-order transgressive–regressive (T–R) or deepening–shallowing units for interbasinal correlation of Permo-Carboniferous rocks of Pennsylvania, Ohio, and West Virginia and the western midcontinent. T–R units were defined using lithostratigraphic and paleoecologic data. In their scheme, if the number of such units between reliable marker beds was

FIGURE 6.4 Changes in eccentricity, tilt, and precession for the last 250,000 years. Addition of these curves for each point in time results in a composite represented by the oxygen isotope curve (figure 6.3). *Modified from Imbrie and Imbrie, 1979.*

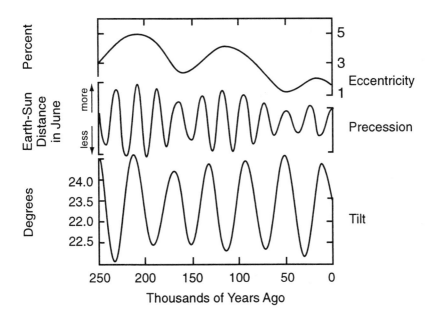

FIGURE 6.5 Effect of precession and obliquity for two extreme conditions (ice growth and ice melt). In the ice growth configuration, when the northern hemisphere is tilted toward the Sun (summer), it is also farthest away, and when the earth is closest to the Sun, the earth is tilted away (winter), so that incoming solar radiation is minimized. In the ice melt configuration, summer in the northern hemisphere occurs when the earth is closest to the Sun, and winter when it is farthest away; hence, ice growth is maximized. Note that in either configuration cyclic changes in the angle of the earth's axis of rotation (precession or wobble) are also superimposed on the cycle of obliquity. The extreme eccentricity of the earth's orbit (greatly exaggerated in this diagram) accentuates climatic extremes. When the orbit is essentially circular, the climatic extremes are more attenuated. Compare with figures 6.3 and 6.4. *From T. H. van Andel,* New Views on an Old Planet. © *Cambridge University Press 1985, 1994. Reprinted with the permission of Cambridge University Press.*

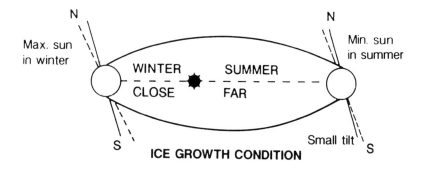

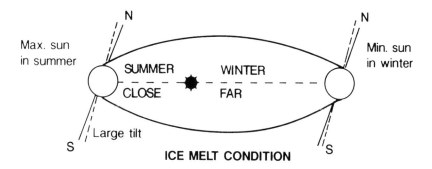

the same from one locality to the next, it was inferred that all were correlative and allocyclic (driven by external forces such as eustatic sea-level fluctuations). If the number of T–R units varied from site to site and none were correlative between marker beds, then the units were regarded as autocyclic (driven by local processes such as delta lobe switching). Most stratigraphic sequences were a mixture of allocyclic and autocyclic units, but autocyclic units were identified as noncorrelative ("extra") units relative to correlative units. They thought that *limited* intrabasinal correlation of sixth-order units (punctuated aggradational cycles or PACS of Goodwin and Anderson, 1985) might also be possible using this framework. According to Busch and West (1987), the T–R framework could be verified by using biostratigraphic, magnetostratigraphic, and chemostratigraphic (such as volcanic ash or soil horizon) datums, but the T–R framework itself provided higher temporal resolution than these other datums, and served as the basis for detailed paleogeographic reconstructions.

CONTINENTAL CONFIGURATION AND CLIMATE[1]

Sea level is not the only thing that goes up and down on this planet, of course. So do the arms of convection cells in the earth's atmosphere caused by differential heating of the earth's surface. Consider atmospheric circulation on a nonrotating earth. At the equator, sunlight has a thinner atmosphere to penetrate than at higher latitudes, where the Sun's rays enter at an oblique angle to the atmosphere. Consequently, fewer rays are reflected away from the earth near the equator, which is why low latitudes tend to be warm. Warm air rises because it is less dense, and as it rises it cools, releasing its moisture as rain (so the tropics are associated with wet, lush rain forests). But the air only rises so far. Eventually the density of the rising air equals that of the surrounding atmosphere and it begins to move laterally toward both poles. This air also cools along its path, and it eventually begins to sink at around 30° north and south latitude, and as it does so it soaks up moisture (as in the Sahara Desert). As they near the earth's surface, each of these descending air masses splits in two, with one-half moving to the north and the other to the south, so the pattern repeats itself at higher latitudes.

Rotation of the earth imparts an effect called the Coriolis force to air flow in the atmosphere. A point on the earth's surface at the North

[1]Much of this section is based on van Andel (1994).

143

Pole moves slower than one at the equator because in the same unit of time, the point at the equator must travel a greater distance. If the point at the North Pole moves toward the equator, maintaining its original velocity (because of the point's inertia), it will lag behind the one at the equator, imparting a curvature to its path. Conversely, a point moving from the equator toward the North Pole moves ahead of the point at the pole. A similar phenomenon occurs in the southern hemisphere, but is the mirror image of that in the northern hemisphere.

The curvature of the air flow establishes the major patterns of atmospheric circulation. As air descends at 30° north and south latitude, it tends to move from east to west over the earth's surface (see Tarbuck and Lutgens, 1985, their figure 14.13). These are the famous Trade Winds (or just Trades) that mariners have used over the centuries. Warm air rising at about 60° north and south latitude tends to flow from west to east because of the Coriolis effect; these winds are the well-known Westerlies that move major weather systems over the United States and Europe. Another set of winds, the Polar Easterlies, lie further north and move from east to west, like the Trades. As these wind systems move over the surface of the oceans, they push the water ahead of them (because of friction), which produces the major oceanic surface currents (figure 6.6). The circular currents are known as gyres and are the result of the Coriolis force. Ancient mariners also learned about the predominant wind and current directions and used them on their return trips to and from the Americas.

The situation is more complicated if land masses are added to the earth's surface. If one gigantic supercontinent (such as Pangaea) exists, there is also one gigantic superocean (Panthalassa; *thálassa* is Greek for "sea"). As surface water moves from east to west in the superocean near the equator, it tends to warm because it is exposed to sunlight along its journey, so the western part of the equatorial ocean is warm (figure 6.7).

Other climatic extremes were associated with Pangaea. The supercontinent may have exhibited megamonsoons, which were much larger versions of the massive air flows found today over India and northwestern Africa (Parrish, 1993). Also, sea level was quite low by Phanerozoic standards (it appears to have been slightly lower than today, which gives an idea of how high sea level was during much of the rest of the Phanerozoic; figure 6.1), so the marine record is much less complete than at other times. Much of the interior of the supercontinent was characterized by arid conditions and interior drainage, and nonmarine sediments, such as giant cross-bedded sands, were common during this time. The earth's albedo (reflectivity) presumably also increased, helping to cool the planet (land reflects sunlight away from the earth, whereas oceans tend to absorb it as heat). Glaciers also characterized much of the late Paleozoic in the southern hemisphere, which would have stimulated ocean circulation rates. These overall conditions of low sea level, cool, dry climate, and rapid ocean circulation rates have been called oligotaxic conditions (table 6.2) by Fischer and Arthur (1977), because it is during these times that biotic diversity tended to be lowest. The term *oligotaxic* and its antonym *polytaxic* (referring to high diversity) were originally applied to much shorter intervals of time, a few tens of millions of years during the Meso-Cenozoic, by Fischer and Arthur, but they appear to be equally descriptive of much longer intervals of time.

If the supercontinent begins to rift apart, separate ocean basins and seaways become established (figure 6.8). During the Mesozoic, for example, the equatorial Tethyan Seaway (after Tethys—pronounced with a long *e*—a daughter of Uranus and Gaea, and wife of Oceanus and mother of the Oceanids and river gods) flowed around the earth largely unimpeded. A similar but by no means identical situation occurred

FIGURE 6.6 Major current patterns of the modern oceans. The large circular patterns are called gyres and result from the Coriolis force. *From T. H. van Andel*, New Views on an Old Planet. © *Cambridge University Press 1985, 1994. Reprinted with the permission of Cambridge University Press.*

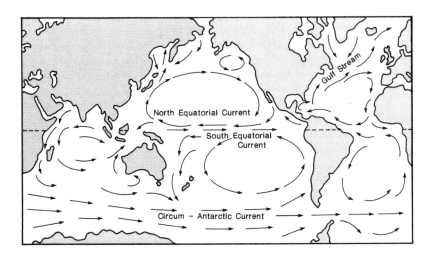

FIGURE 6.7 Effects of a supercontinent on ocean circulation and global temperature. *From T. H. van Andel,* New Views on an Old Planet. © *Cambridge University Press 1985, 1994. Reprinted with the permission of Cambridge University Press.*

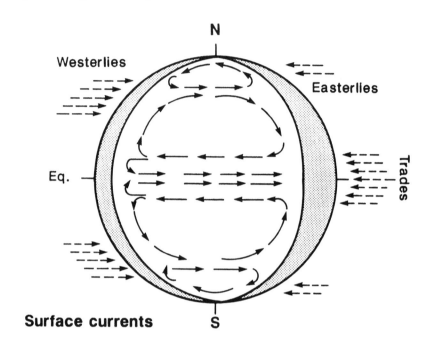

Surface currents

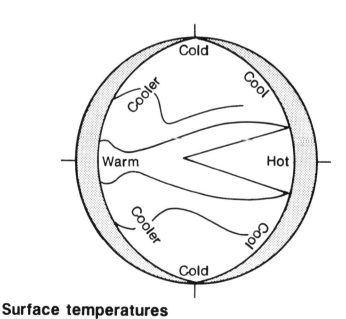

Surface temperatures

during most of the first half of the Paleozoic (Cambrian to Devonian), when the Precambrian supercontinent Rodinia (taken from the Russian *rodit'* by McMenamin and McMenamin, 1990, meaning to "to beget" or "to grow") rifted apart and the proto-Atlantic Ocean called the Iapetus Ocean (after Iapetus, a son of Oceanus and Gaea, a brother of Tethys, and father of Atlas and Prometheus) existed between what was then North America (Laurentia) and Europe (Baltica). These times were characterized by warm, extensive, shallow seas several hundred meters or more higher than today's, so the earth's albedo was presumably somewhat lower. These are prime conditions for the formation of carbonate (limestone) platforms, in which reefs—formed by various taxa through time—thrived (see James, 1983). Much of the world's petroleum comes from rocks of the Cambrian-to-Devonian and Mesozoic (Tissot, 1979); these intervals are also characterized by widespread black shales, which has led some workers to suggest that ocean circulation was quite sluggish and conditions highly reducing (anoxic, or without oxygen). Such conditions would have allowed preservation

TABLE 6.2 Phenomena Associated with Icehouse and Greenhouse States of the World

Icehouse State (example: present earth)	Greenhouse State (example: Cretaceous oceans)
Ocean is highly stratified.	Much less stably stratified than icehouse state
Very stable; recycling of nutrient-rich deep water is difficult.	Surface water temperatures not much higher than icehouse state at equator, but 12–15°C in high latitudes.
Surface water temperatures range from <2°C (circumpolar) to >25°C (equatorial).	Deep water temperatures from 15°C (equatorial) to about 10°C (circumpolar).
Bottom water ranges from about +2°C (interglacial) to +1°C (glacial).	Low-density surface water leads to slow bottom water flow; water at 15°C holds half the oxygen of 2°C water, hence little oxidative power.
Vigorous flow of bottom water, rich in oxygen, hence strong oxidation, little storage of organic matter.	Little recycling of organic matter, much buried in sediment; nutrient recycling much reduced.
Environments diverse; high productivity in areas of upwelling.	Low productivity, but good oil source beds.

Modified from van Andel, 1994.

147

FIGURE 6.8 Highly generalized ocean circulation as a supercontinent rifts apart and continents move over poles. Compare Circum-Antarctic Current in figure 6.6. *From T. H. van Andel,* New Views on an Old Planet. © *Cambridge University Press 1985, 1994. Reprinted with the permission of Cambridge University Press.*

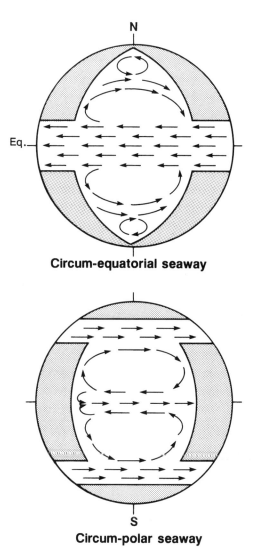

Circum-equatorial seaway

Circum-polar seaway

of large amounts of organic matter in sediments, including petroleum (see chapter 7).

What if the Tethyan Seaway had been blocked by a land mass? This is precisely what began to happen in the Cenozoic with the collisions of Africa and Europe (to form the Alps) and India and Asia (to form the Himalayas), although the African plate moved away from Europe later. Also, approximately 3 million years ago, the Isthmus of Panama breached the surface of the ocean (through tectonic uplift) and separated the equatorial Atlantic from the Pacific. Warm water, and the associated warm, moisture-laden air, were diverted into the Gulf of Mexico, where they basically did a U-turn (Florida Loop Current), flowed back out around the Florida Keys through the Straits of Florida, and then northward, between Florida and the Bahamas, where they headed out to sea off Cape Hatteras to help form the Gulf Stream (figure 6.6). These waters moved moisture to high latitudes, where it began to condense (this is why the British Isles are foggy), and only about 200,000 years later (2.4 to 2.8 Ma), deep-sea cores taken off the British Isles exhibit evidence of ice-rafted debris. The "Ice Ages" had begun. Some workers have suggested that the isolation of Antarctica over the South Pole by ocean currents (Circum-Antarctic Current) beginning about the Late Eocene to Oligocene may have set the stage (a possible constraint) for the subsequent advance of northern hemisphere ice sheets by initially cooling the earth.

CYCLES OF CO_2

Insolation and glaciers, by themselves, are not the whole story of the Ice Ages, however. Over the past two decades or so, continuous records of atmospheric composition have been obtained from ice cores taken in Greenland and Antarctica. Bubbles of air are trapped in the ice as it forms and so constitute a record of atmospheric composition, which is obtained by melting thin slices of ice (and releasing the "fossil air" trapped in bubbles) under controlled laboratory conditions. The ice core records indicate that atmospheric CO_2 has fluctuated in concert with the oxygen isotope record, from about 200 ppm during glacial intervals (such as oxygen isotope stage 2; figure 6.3) to about 280 ppm in the Holocene. At least, that is what it should be; it is actually more than 350 ppm, based on continuous records taken at stations in Hawaii and elsewhere that indicate a steady rise over the last few decades, presumably because of fossil fuel combustion (see Sundquist, 1985, figure 1). Nevertheless, the amount of CO_2 pumped into the atmosphere per year by burning of fossil fuels is substantially lower (by about 0.3–0.4 gigatons per year) than expected, and some workers have suggested that terrestrial vegetation and soils may act as the missing "sink" (Fisher et al., 1994; see also Davidson et al., 1995a). Elevated atmospheric CO_2 levels may even stimulate photosynthesis ("CO_2 fertilization"), thereby alleviating some of the anthropogenic greenhouse effect, but faster-reproducing species may outcompete slower-reproducing ones, so that terrestrial plant (and therefore animal) diversity may decline (Bazzaz and Fajer, 1992; Monastersky, 1995).

Numerous scientists have attempted to explain what regulates these changes in atmospheric CO_2. Most studies have focused on the role of the oceans because modern oceans circulate, on average, once about every thousand years (based on the distribution of radiocarbon in ocean water left over from atomic weapons tests). Presumably, the oceans have exhibited similar circulation rates at least for the last several million years through the Plio-Pleistocene, when northern hemisphere ice caps first began to develop and deep-ocean circulation became more and more like today's. The basic idea is that somehow the deep-ocean currents change so that phytoplankton are stimulated to photosynthesize more and therefore sequester more CO_2 into organic matter (CH_2O), some of which sinks to the ocean bottom and is stored there in sediment:

$$CO_2 + H_2O \rightarrow (CH_2O)_n\downarrow + O_2, \qquad (6.3)$$

where CH_2O is a generic form of organic matter (for example, if $n = 3$, then the product is $C_6H_{12}O_6$, or glucose).

In one of the more elaborate schemes involving marine photosynthesis, Broecker and Peng (1989) advanced the polar alkalinity hypothesis. This scheme is based on the fact that two basic deep-water masses are produced in the Atlantic Ocean basin: Antarctic bottom water (AABW) and North Atlantic deep water (NADW). AABW actually flows over the floors of ocean basins all over the world and is an "old" water mass. In doing so, AABW picks up a great deal of CO_2 from animal respiration and organic matter decay, and the nutrients that are released by decay. Carbon dioxide dissolves into seawater via the equation

$$CO_2 + H_2O + CO_3^{2-} \rightarrow 2HCO_3^- \qquad (6.4)$$

(Broecker and Peng, 1982). Because AABW is CO_2 rich (and therefore CO_3^{2-} rich; see equation 6.2), it can store large amounts of CO_2. Also, it is nutrient rich.

In contrast, NADW, which is produced off the coast of Greenland in the North Atlantic, is a "young" water mass. Its formation probably began in earnest following the closure of the Straits of Panama and today its flow is about equal to that of 100 Amazon Rivers, or about 20 million cm^3 sec^{-1} (Broecker, 1995). NADW is CO_2 poor and nutrient poor because it flows to the south only through the Atlantic basin. NADW eventually rises toward the surface (upwelling) to replace AABW, which is sinking off Antarctica (figure 6.9). As it does so, it

FIGURE 6.9 Distribution of Antarctic bottom water (AABW) and North Atlantic deep water (NADW) during interglacial times. According to the polar alkalinity hypothesis, NADW production slows during glacials, thereby resulting in lowered atmospheric CO_2. *From T. H. van Andel, New Views on an Old Planet. © Cambridge University Press 1985, 1994. Reprinted with the permission of Cambridge University Press.*

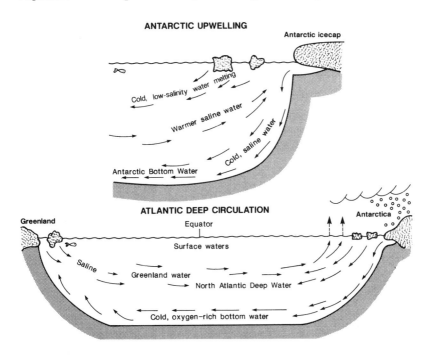

151

mixes with AABW and dilutes its nutrient and CO_3^{2-} content, making AABW less able to hold CO_2. Consequently, atmospheric pCO_2 (partial pressure or concentration of CO_2) rises.

According to the polar alkalinity hypothesis, NADW production tends to slow during glacials. This is indicated by Cd/Ca ratios, which are used as a proxy for phosphorus (a limiting nutrient on geological timescales) in the tests of foraminifera (Boyle and Keigwin, 1982, 1987). Ice sheets that advance over the North Atlantic hinder the formation of NADW production; also, NADW does not flow as far into the Atlantic basin as during interglacials (including today). Consequently, AABW is no longer being diluted; it is able to incorporate more CO_2 into organic matter (through photosynthesis by marine plankton, especially siliceous forms called diatoms) and HCO_3^- and atmospheric pCO_2 decrease.

There is evidence, from carbon isotopes, that such an inverse relationship between NADW and AABW production occurs. Like oxygen, carbon exists in two stable forms: ^{13}C and ^{12}C (^{14}C is radioactive). The lighter (^{12}C) isotope is preferentially incorporated into organic matter during photosynthesis, while leaving behind ^{13}C in seawater (positive shift or increased ratio of $^{13}C/^{12}C$ or $\delta^{13}C$). The heavier isotope is thus incorporated into the shells of marine plankton (foraminifera) when marine photosynthesis increases. On the other hand, if marine photosynthesis decreases, $\delta^{13}C$ also decreases and is said to undergo a negative shift ($\delta^{13}C$ also changes according to changes in the rate of terrestrial photosynthesis; see chapter 7). The ratio of the two isotopes ($\delta^{13}C$) is most often measured on the shells of plankton to gauge ancient productivity. But it can also be measured on benthic foraminifera as a measure of dissolved CO_2 content of bottom water masses; if dissolved CO_2 goes up, then $\delta^{13}C$ goes down. Evidence from deep-sea cores taken off Antarctica and in the North Atlantic indicates that $\delta^{13}C$ does in fact decrease during glacials (suggesting decreased NADW production) and increases during interglacials (increased NADW production; see Oppo and Fairbanks, 1987; Oppo and Lehman, 1995). This confirms the influence of changes in NADW production on AABW chemistry suggested by the polar alkalinity hypothesis.

A number of the negative $\delta^{13}C$ shifts correspond to the deposition of ice-rafted debris in North Atlantic deep-sea cores. These Heinrich events are consistently preceded by the appearance of cold-water planktic foraminiferal assemblages (Maslin et al., 1995), which suggests that ice sheets were expanding over the North Atlantic (Bond, 1995; Bond and Lotti, 1995). Glacial intervals are also marked by finer-grained sediments, suggesting that slowing of NADW production caused less bottom scour (McCave et al., 1995). Based on computer models, Rahmstorf (1995) found that NADW production is highly sensitive to fresh (melt) water input that can cause temperature changes of a few degrees within a matter of years.

The polar alkalinity hypothesis makes sense given the rapid circulation rate (about 1000 years) of the oceans because the rapid turnover time of the oceans could account for rapid climate change. But despite the attractiveness of Broecker's hypothesis, this mechanism appears to be incapable of drawing down sufficient quantities of CO_2 to cool the earth. Also, later studies found no direct evidence for enhanced productivity in the waters off Antarctica (see Singer and Shemesh, 1995). Nevertheless, other workers have found high nutrient levels, particularly nitrate, in cores taken through glacial sediments elsewhere, such as in the Arabian sea (Altabet et al., 1995) and off the coast of Mexico (Ganeshram et al., 1995). Today these areas are sites of denitrification by sulfate-reducing bacteria, in which nitrate (NO_3^-) is reduced to forms unusable by plankton (nitrite: NO_2^-; nitrous oxide: N_2O; nitrogen: N_2). Nitrogen is yet another element that exists in stable isotopic forms: [14]N and [15]N. Denitrifying bacteria preferentially use the lighter isotope during denitrification, leaving behind [15]N. Like oxygen isotopes, the ratio of the two is measured and then used as an indicator of relative rates of denitrification and nitrate availability. Altabet et al. (1995) found that changes in $\delta^{15}N$ were synchronous with Milankovitch cycles, which may have affected the monsoon over the Arabian peninsula and rates of coastal upwelling.

Most recently, Broecker (1997) has implicated El Niño in global climate change. He has suggested that changes in atmospheric CO_2 and water vapor (which is the earth's dominant greenhouse gas) are both involved in altering the earth's temperature: as CO_2 rises, the atmosphere can hold more water vapor, which accentuates the warming effect. These changes may occur via changes in upwelling rates (and nutrient levels?) in the eastern equatorial Pacific, which have been implicated in El Niño. In turn, changes in upwelling rates implicate reorganization of ocean circulation (as evidenced by varved sediments from the Santa Barbara basin), which again hints that the ocean's conveyor belt is involved.

Whatever is going on, it appears that the earth has had more than one quasistable mode of operation during the Pleistocene (Broecker et al., 1985; Broecker, 1997). Whatever the control on greenhouse gases, it appears that changes in ocean circulation are intimately involved in moving (and may have even pushed) the earth's climate from one mode to another. These changes may be abrupt (less than 10 years), as the rapid transition from cool to warm intervals in ice records indicate, and once the climate shifts have occurred, they appear to last for up to 1000 years (Broecker, 1995). After rapid cooling (less than 10 years), summers in Dublin, for example, would resemble those of Spitsbergen (Greenland) and winters in London those of Siberia (Broecker, 1995).

A very unpleasant thought is that even within warm intervals (such as oxygen isotope stage 5, which represents the last major benign episode of the Pleistocene), abrupt climate shifts may occur with apparently little or no external forcing. These conclusions were originally based on recent ice core records from Greenland for a time called the Eemian (about 110,000 to 130,000 Ma, within isotope stage 5). Possibly, the abrupt shifts represent artifacts produced by ice flow at depth, but recent pollen records from central Europe suggest similar rapid climate shifts on land (see Monastersky, 1994, and Zahn, 1994, for brief reviews).

Thus carbon dioxide (from fossil fuels and water vapor) could be accumulating in the earth's atmosphere with little or no effect until a threshold is crossed into an entirely new mode (Broecker, 1987, 1995). In chaos theory, an unstable dynamical system can be moved by a disturbance (no matter how small) from one attractor to another. Does the earth's climate regime even exhibit chaotic behavior? The oxygen isotope curve (figure 6.3) for the last half of the Pleistocene bears a strong resemblance to the chaotic behavior exhibited by the logistic system (figure 5.6d). Chaos has also been detected in what was always thought to be one of the most highly deterministic natural systems known: the solar system. Although not certain, the shift from obliquity-dominated to eccentricity-dominated cycles may have resulted from chaotic orbital behavior (van Andel, 1994:249; see also Laskar, 1989; Krumenaker, 1994).

There are also long-term controls on atmospheric CO_2. In Fischer's cycles (figure 6.1), seafloor spreading rates increased volcanism at spreading centers and subduction zones. According to Fischer, increased atmospheric pCO_2 from volcanism presumably caused the earth's average surface temperature to rise during the Cambrian-to-Devonian and Jurassic-to-Late Eocene (greenhouses). Just the opposite conditions supposedly prevailed during icehouses (Late Devonian–to-Triassic and Oligocene-to-Recent) as a result of decreased spreading rates and MOR volume.

These broad changes in the earth's climate (figure 6.1) were paralleled by the mineralogy of abiotic carbonate cements and ooids. Deposition of predominantly low-Mg calcite cements and ooids during greenhouses presumably reflects the lowered $CaCO_3$ saturation state of surface oceans caused by elevated atmospheric pCO_2. During much of the Cambro-Devonian, for example, atmospheric pCO_2 was apparently quite high, about 10–20 times those of modern values based on Berner's (1990, 1993), models; $CaCO_3$ saturation of surface waters was presumably quite low as a result (Mackenzie and Morse, 1992) and diagenesis of sediments was enhanced (Bramlette, 1958; Walker and Diehl, 1985). Diagenesis may also have been enhanced by sulfate-reducing bacteria when anoxia was widespread (Vasconcelos et al., 1995), as it apparently was during greenhouses (see chapter 7). Conversely, high-Mg calcite and aragonite cements and ooids dominated during icehouses, when CO_2 levels were apparently lower (Mackenzie and Pigott, 1981; Sandberg, 1983; Mackenzie and Agegian, 1989; Wilkinson and Given, 1986; Wilkinson and Algeo, 1989; Morse and Mackenzie, 1990; Mackenzie and Morse, 1992; see also Riding, 1993). Bates and Brand (1990) suggested that this cyclic variation in ooids and cements was primarily the result of diagenesis. According to them, aragonitic ooids and cements were dominant throughout the Phanerozoic, but were altered to calcite when CO_2 levels were high; later studies appear to have largely confirmed the broad relationship between CO_2 and primary mineralogy (Morse et al., 1997).

The mineralogy of microfossil assemblages may shed some light on the matter, as well as on a long-standing debate in paleobiology: the origin of skeletons (see also chapter 5). Why did skeletons appear in the first place? The causal mechanisms include increased body size (and therefore the need for support). This in turn may have resulted from an increase in atmospheric oxygen (Canfield and Teske, 1996). Indeed, Logan et al. (1995) suggested that increased production of fecal pellets in the late Precambrian sent larger amounts of organic matter to the ocean bottom, rather than allowing it to decay and use up O_2; before this time, bacterial degradation of organic matter was common. Also, detoxification of excess Ca^{2+} ions, which are toxic at high concentrations, may have occurred by secretion into hardparts. Finally, protection may have been needed from predation. In reviewing these hypotheses, Vermeij (1989) comes down strongly on the side of predation (he is also a biologist). Similarly, Culver (1991) suggested that the foraminiferal test may have arisen as a protective response to indiscriminate scavenging and predation by deposit feeders in the Precambrian.

This may certainly be the case, but the subsequent record of test mineralogy of foraminiferal suborders also appears to reflect biomineralization in response to ocean chemistry (figure 6.10). Early tests consist of organic matter or sediment grains agglutinated together with organic cement, but, curiously, the majority of calcareous suborders of foraminifera appeared during times characterized by increasing $CaCO_3$ saturation of ocean surface waters (greenhouse-to-icehouse transitions or full icehouse phases). Moreover, the strong similarity in ultrastructure of abiotic cements and the test of certain foraminiferal suborders (Fusulinina, Miliolina) suggests a primary response to $CaCO_3$ saturation of surface waters by these organisms, not secondary alteration (Martin, 1995a). In other words, test mineralogy of foraminiferal suborders tends to recapitulate ocean chemistry ($CaCO_3$ saturation). Other workers have also noted a correspondence between hardpart mineralogy of macrofossils and seawater chemistry of the late Paleozoic to Triassic (Wilkinson, 1979; Railsback and Anderson, 1987; Railsback, 1993; Van de Poel and Schlager, 1994).

Martin (1995a) suggested that foraminifera may have responded to Ca^{2+} toxicity (related to $CaCO_3$ saturation) by establishing symbiotic (mutualistic or mutually beneficial) associations with bacteria capable of secreting $CaCO_3$. $CaCO_3$-secreting organelles that may contain their own DNA have been found in some species of foraminifera, and suggest that these organelles were originally $CaCO_3$-secreting bacteria eaten by the ancestral foraminifers (West, 1995). Instead of being digested, the bacteria persisted in their hosts' cytoplasm and came to help regulate $CaCO_3$ secretion. A similar scenario of endosymbiosis was proposed by Margulis (1970) and earlier workers (Mereschkovsky, 1910; Wallin, 1923) for the origins of the cell organelles called mitochondria (where the chemical reactions of cellular respiration occur) and chloroplasts (where photosynthesis occurs), both of which also have their own DNA (see Dyer and Obar, 1994, for further evidence and review of the field). This theory of symbiogenesis initially met a great deal of resistance and was widely scoffed at, but over the years the symbiotic origin of mitochondria and chloroplasts has become widely accepted.

The composition of fossil plankton assemblages also appears to reflect the $CaCO_3$ saturation of the oceans during the Paleozoic. Based on examination of ophiolite complexes through the Phanerozoic, Boss and Wilkinson (1991) concluded that deposition of extensive cratonic limestones probably helped keep the carbonate compensation depth (CCD) between 1 and 2 km depth, thereby preventing significant deep-

FIGURE 6.10 Geologic distribution and composition of major microfossil taxa. Thickness of line indicates relative species diversity (period-averaged data). Note that Cambro-Ordovician microfossil assemblages are dominated by noncalcareous taxa (organic and agglutinated suborders of benthic foraminifera, acritarchs, graptoloids, radiolaria, and conodonts). Most calcareous suborders of foraminifera appeared later, during greenhouse-to-icehouse transitions (Late Ordovician–Early Silurian; Late Devonian–Early Carboniferous) or full icehouse phases (Permo-Triassic), when $CaCO_3$ saturation of the oceans was presumably high. In most cases, the mineralogy, and in some cases ultrastructure of test crystallites (Fusulinina, Textulariina, Miliolina, Carterinina), reflects $CaCO_3$ saturation levels of surface oceans when suborders appeared. All agglutinated foraminifera are placed in the suborder Textulariina for simplicity. Question marks denote enigmatic occurrences. C = short icehouse episode, F = foraminiferal suborder, LMG = low-Mg calcite, HMG = high-Mg calcite, ARAG = aragonitic. Greenhouse–icehouse cycles, warm–cool modes, and mineralogy of abiotic cements and ooids as in figure 6.1. See text for further discussion. *Reprinted from* Global and Planetary Change, *volume 11, Martin, R. E., Cyclic and secular variation in microfossil biomineralization: Clues to the biogeochemical evolution of Phanerozoic oceans, pp. 1–23, 1995, with kind permission of Elsevier Science–NL, Sara Burgerhartstraat 25, 1055 KV Amsterdam, The Netherlands.*

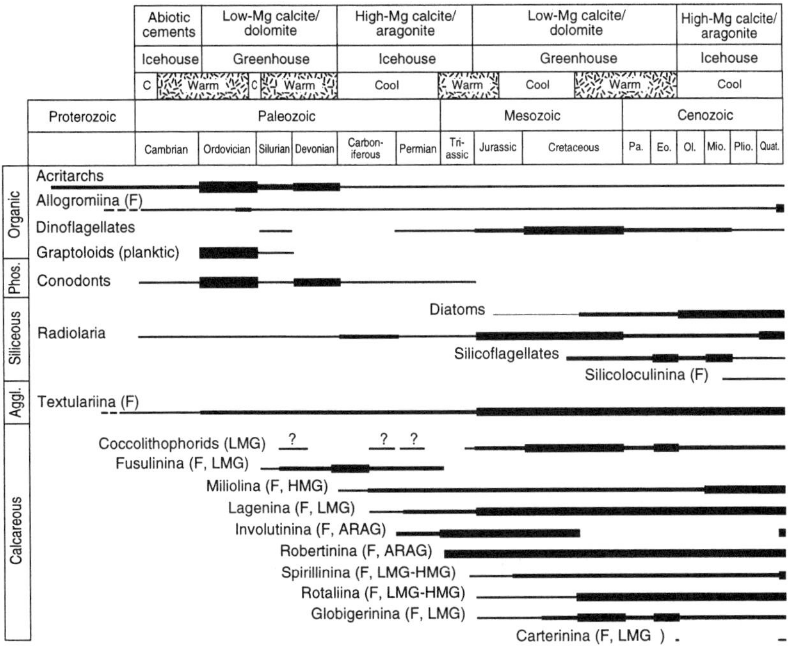

sea ooze formation. The entire ocean is basically a gigantic carbonate buffer system, which maintains a pH of about 8.1–8.3 when in equilibrium with atmospheric CO_2. If CO_2 levels rise or too much limestone is deposited on continental shelves (as it was in the Cambro-Devonian and much of the Mesozoic), the deep-sea is starved for $CaCO_3$ and the CCD rises to dissolve sufficient $CaCO_3$ in sediments to maintain the ocean's pH. If cratonic limestones are eroded or CO_2 falls, the CCD deepens accordingly. Thus the CCD has probably moved up and down cyclically in the ocean basins through the Phanerozoic (van Andel, 1975; Moore et al., 1982). In the case of the Cambro-Devonian, it appears that if calcareous plankton did exist in the Paleozoic (coccolithophorids and planktic foraminifera; figure 6.10), they must have dissolved, leaving behind dissolution-resistant noncalcareous taxa such as acritarchs, graptolites, and radiolaria (figure 6.10).

EVOLUTION OF THE LONG-TERM CARBON CYCLE

Despite the heavy emphasis in this chapter on the role of physical mechanisms in climate control, life has not merely been along for the ride. Stromatolites (colonial cyanobacteria or blue-green algae), for example, were responsible for oxygenating the earth's atmosphere once suitable habitats (continental shelves) were present in sufficient abundance; the initial oxygenation may have led to the formation of the economically important banded iron formations. Iron in its reduced or ferrous state (Fe^{2+}) is highly soluble in water, but when oxidized to its ferric state (Fe^{3+}, or "ferric the red") it is highly insoluble. As the surface waters of the oceans became increasingly oxygenated as a result of stromatolite photosynthesis, iron probably began to precipitate in significant quantities.

The calcareous plankton (coccolithophorids and planktic foraminifera) have been no less important in controlling the composition of the earth's atmosphere. Chemical weathering occurs when atmospheric CO_2 (carbon dioxide) dissolved in rain and groundwater attacks rocks to release Ca^{2+} and form bicarbonate ions (HCO_3^-), which are carried in dissolved form by rivers to the oceans:

$$2CO_2 + H_2O + CaSiO_3 \rightarrow Ca^{2+} + 2HCO_3^- + SiO_2. \quad (6.5)$$

(For the sake of simplicity, I use $CaSiO_3$, which is actually the mineral wollastonite, for "granite"; see Berner and Lasaga, 1989. Other types of rocks, igneous and sedimentary, may be weathered in a similar manner in simulations; see Berner et al., 1983; Berner, 1990, 1993). This portion of the cycle acts as a negative feedback when atmospheric CO_2 becomes elevated. In marine waters, the ions from runoff are incorporated into the calcareous shells ($CaCO_3$) of marine plankton:

$$2HCO_3^- + Ca^{2+} \rightarrow CaCO_3\downarrow + CO_2 + H_2O. \qquad (6.6)$$

Eventually the plankton die and a significant proportion of the shells settle to the bottom before dissolution. Over intervals of millions to tens of millions of years, the calcareous ooze that is formed is eventually carried by spreading ocean crust to oceanic trenches, where it is subducted, melted, and the CO_2 present in the skeletons is released back to the atmosphere, thereby completing the cycle:

$$CaCO_3 + SiO_2 \rightarrow CaSiO_3 + CO_2. \qquad (6.7)$$

Uplift, erosion, and oxidation of carbon-rich rocks (such as limestones) may have also contributed to global warming beginning in the Paleocene (see Beck et al., 1995). Without the advent of this portion of the cycle (equation 6.7), today's climate might be substantially cooler (Volk, 1989a).

Negative feedback from continental weathering is supported by another set of stable isotopes. Although the initial ratio of ^{87}Sr to ^{86}Sr is the result of radioactive decay of rubidium in igneous rocks, thereafter $^{87}Sr/^{86}Sr$ ratios behave like stable isotopes (Holser et al., 1988). The $^{87}Sr/^{86}Sr$ ratio reflects primarily a balance between weathering of high-$^{87}Sr/^{86}Sr$ continental rocks and input of low-$^{87}Sr/^{86}Sr$ via hydrothermal exchange between ocean crust and seawater (Raymo, 1991; Richter et al., 1992; Ingram et al., 1994; Jones et al., 1994). Raymo et al. (1988) and Raymo (1994) attributed the overall rise of the $^{87}Sr/^{86}Sr$ curve during the late Cenozoic to mountain building and dissolved riverine fluxes to the oceans. Similarly, increased $^{87}Sr/^{86}Sr$ ratios in the Late Precambrian, Late Ordovician, Late Devonian, and Pennsylvanian-to-Permian (except for the latest Permian; figure 6.11) suggest increased rates of continental weathering (and CO_2 drawdown) as a result of orogeny in North America and Europe. These intervals also correspond to intervals of southern hemisphere glaciation and sea-level fall (figure 6.11; cf. figures 6.1, 6.10; note the appearance of the foraminiferal suborder Fusulinina in the Late Ordovician). These glaciations sometimes occurred when atmospheric CO_2 levels were normally quite high. This suggests that continental weathering and continent–ocean configuration (movement of a continent over a pole where glaciers can develop) are able to counteract greenhouse conditions (see Crowley and Baum, 1991). Some workers have recently begun to use strontium isotope ratios in high-resolution correlation of rocks, although there is still error of 1–3 million years (see Miller et al., 1988; Smalley et al., 1994; chapter 2).

Based on the fossil record, calcareous plankton (planktic foraminifera and calcareous nannofossils such as coccolithophorids) presumably did not evolve until the Mesozoic (figure 6.10), so their integration into the long-term global carbon cycle could not have begun until the Mesozoic. Although this appears to be true for planktic foraminifera, enigmatic calcareous nannofossils have been reported from cratonic sediments of the Paleozoic (figure 6.10; see Tappan, 1980; Jafar, 1983; Boss and Wilkinson, 1991, for references), although they have typically been dismissed either as contaminants from younger rocks or for lack of adequate description (see Lord and Hamilton, 1982). If taken seriously, though, these microfossil Lagerstätten suggest a much longer history for the group. If calcareous plankton were present in the Paleozoic, most of them must have dissolved because of greater rates of diagenesis, as discussed earlier.

FIGURE 6.11 Strontium isotope ($^{87}Sr/^{86}Sr$) ratios through the Phanerozoic. Black bars indicate times of apparent glaciation. Major orogenic episodes are also indicated. *From M. E. Raymo, 1991. Geochemical evidence supporting T. C. Chamberlin's theory of glaciation.* Geology 19:344–347. *Reprinted with permission of the Geological Society of America.*

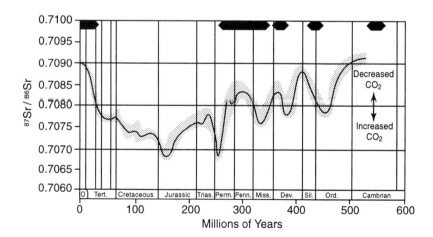

But why did calcareous plankton apparently originate in shallow seaways and only later begin to populate the open ocean in the Mesozoic? Here we must turn to other fossil evidence. Siliceous oozes (today composed mainly of diatoms and radiolaria) preceded calcareous ones by several hundred million years (figure 6.10), possibly because the use of SiO_2 for shell construction is energetically more efficient than $CaCO_3$ (Brasier, 1986:244). Interestingly, the locus of siliceous ooze deposition shifted across continental shelves during the Phanerozoic. Maliva et al. (1989) concluded that late Proterozoic cherts were mainly peritidal and of abiological origin (based on lack of petrographic evidence for skeletal sources of silica). By the Cambrian, biogenic cherts were predominantly shallow subtidal in occurrence and the byproduct of evolutionary diversification of demosponges and problematic siliceous scale-producing protists (Maliva et al., 1989). After the rise of radiolarians in the Cambro-Ordovician, the locus of siliceous ooze deposition shifted across the shelf to the open ocean (Maliva et al., 1989; Casey, 1993). In the Mesozoic, radiolarians began to give way to silicoflagellates (which prefer moderate to high nutrient levels; Lipps and McCartney, 1993), and by the Miocene, to diatoms (figure 6.10; Tappan, 1980), which today prefer nutrient-rich (eutrophic) conditions (Kilham and Kilham, 1980). Today, diatoms account for about 20–25 percent of total global primary production (photosynthesis; Werner, 1977). In fact, they secrete so much silica that only about 3 percent accumulates in siliceous oozes, the rest being dissolved in the upper water column and recycled back to the surface, where it is used again (Tréguer et al., 1995). Indeed, diatoms use silica so rapaciously that Harper and Knoll (1975) suggested that they are the primary reason radiolarians came to produce latticelike skeletons in the Cenozoic, as opposed to the more robust shells of their Paleozoic counterparts (Casey, 1993).

The oldest reliable record of diatoms is the Jurassic (Toarcian; Barron, 1993). Of potential significance in this respect is the report of spindle-shaped (pennate?) diatomlike structures from late Proterozoic stromatolitic carbonates (Licari, 1978, their plate 3, figure 9). If diatoms existed before the Jurassic, their apparent absence from open oceans during the Paleozoic and Early Mesozoic may indicate their restriction to nearshore, highly nutrient-rich habitats, in which fragile siliceous eukaryotes (like diatoms) are unlikely to be preserved (Margulis et al., 1980).

As mentioned in chapter 5, the three major evolutionary faunas of the Phanerozoic (Cambrian trilobite-rich, Paleozoic brachiopod-rich, and Meso-Cenozoic mollusc-rich) exhibited a similar onshore–offshore pattern of development. The typical explanation for these patterns revolves around biological specialization during the exploitation of new habitats and niches. This is no doubt part of the explanation, but could these patterns have also been related to the concentration of important nutrients such as phosphorus that in turn constrain rates of marine photosynthesis and the availability of food (energy) to higher levels of food chains? Could the macrobenthos be tracking the plankton?

Another clue is provided by the previous interpretation of the strontium isotope curve. Strontium isotope (^{87}Sr/^{86}Sr) ratios of marine carbonates suggest that significant input of nutrients to the seas from the land occurred through the Phanerozoic. Tardy et al. (1989) used the ^{87}Sr/^{86}Sr curve as a proxy for continental runoff during the Phanerozoic; they found that the ^{87}Sr/^{86}Sr curve is in general agreement with calculated runoff rates (a function of latitudinal position of the continents and relative area of continents and oceans; Tardy et al., 1989). These clues are just part of a larger secular increase (long-term trend through time) in the diversity of the biosphere.

7

Energy and Evolution

You can't step twice into the same river. —*Heraclitus*

Eldredge and Salthe (1984) and Eldredge (1995) argued that there are two biological hierarchies involved in evolution: genealogical and ecological (table 1.1). As might be expected, ultra-Darwinians argue that competition for reproductive success drives competition for resources (food and energy, among them), but Eldredge and Salthe argue that it is the other way around. It is much simpler, according to them, to have the actions of the ecological hierarchy drive evolution of the genealogical one: "All organisms behave economically as a simple consequence of being alive; . . . the structural organization and inner workings of economic systems flow directly from such behavior; . . . economic systems depend on genealogical systems (reproductive behavior) *purely as a constant supply of players in the ecological arena;* and . . . what happens in the ecological arena helps determine the fate of genetic information as it is passed along from generation to generation in the genealogical context" (Eldredge, 1995:196). The connection between the two hierarchies occurs through the shared organismic level (table 1.1). Certainly, the organisms of the ecological hierarchy are constantly being "shuffled" because new kinds of organisms are produced by the genealogical hierarchy (Eldredge and Salthe, 1984). But organisms also require food (energy) to sustain themselves; without energy there would be no search for resources or mates and there would be no evolution. The connection between the two hierarchies is primarily an economic one of food and energy availability (Eldredge and Salthe, 1984; Salthe, 1985).

What would happen if overall food (energy) availability were to increase over geologically long intervals of time? Helen Tappan was among the first to address this issue in detail, but as discussed in chapter 5, many of her assertions conflicted with the fossil record. To my knowledge, however, Tappan was the first scientist to integrate the stable isotope and paleontological records to give a broad view of the history of interaction of the earth's physical systems and biota. Unfortunately, she adhered too closely to a particular paradigm of ecological succession (chapter 5), and the stable isotope curves available to her at the time were crude, so her papers have been treated in recent accounts as being largely of historical interest only. Like Tappan, later workers have examined the fossil record for long-term changes that might reveal the history of the biosphere, but most have ignored the stable isotope record. On the other hand, most geochemists who deal with biogeochemically important elements such as phosphorus and nitrogen, which act as limiting nutrients and are cycled through both the lithosphere and biosphere, have largely ignored the fossil record (Berner and colleagues' models of long-term changes in CO_2 are a major exception). In this chapter, the role of nutrients, marine productivity, and food in the evolution of the biosphere through the Phanerozoic is reassessed, based on more recent stable isotope and lithologic evidence and a reinterpretation of plankton diversity patterns. Rather than being discussed separately, the evidence for each of several major subdivisions of the Phanerozoic is integrated to suggest that there has been a secular increase in nutrient levels and productivity in the oceans.[1] Although many changes are evident in the Phanerozoic fossil record, this one does not stand out unless the entire Phanerozoic is examined (cf. chapter 1).

CAMBRO-DEVONIAN SEAS: PUNCTUATED "SUPEROLIGOTROPHY"

Some of the most oligotrophic (nutrient-poor) waters on this planet today are found in oceanic gyres (see figure 6.6; Worsley et al., 1986). Surface nutrient levels may have been far lower during much of the Cambro-Devonian (superoligotrophic conditions), and therefore unable to sustain large phytoplankton populations. During the Cambro-

---

[1]Much of this discussion is taken from Martin (1995a, 1996a).

Devonian, ice caps occurred only in the Late Ordovician and perhaps Late Devonian (figure 6.11; cf. Frakes et al., 1992); thus ocean circulation during much of this time was presumably driven by the production of highly saline deep-water masses via evaporation in widespread, shallow seaways (Brass et al., 1982; Railsback et al., 1990). This was presumably much like the modern Mediterranean (where surface waters evaporate in the eastern basin, sink, and flow out through the Straits of Gibraltar), but on a much greater spatial scale. Even in modern oceans, though, in which vigorous oceanic circulation is driven by production of deep-water masses near both poles, dissolved PO_4^{3-} levels in the subphotic zone are several times to an order of magnitude or more higher than in the photic layer (Sverdrup et al., 1942; Hallock et al., 1991). Thus during the Cambro-Devonian, when glaciers were largely absent, most nutrients may have been sequestered below the photic zone (Fischer and Arthur, 1977; Sheldon, 1980; Berner and Raiswell, 1983:860–861; Holser et al., 1988). During much of this interval, reduced ocean circulation and widespread anoxia may have resulted in extensive denitrification (by reducing NO_3^- which is used as a nutrient) in the photic zone (Rau et al., 1987) while the high sea level trapped nutrients near the shore (Holser et al., 1988). Also, continental weathering rates were probably too low during much of the Cambro-Devonian to deliver nutrients in the photic zone, as soils were poorly developed (Knoll and James, 1987; Keller and Wood, 1993).

Low rates of marine photosynthesis and organic carbon (C_{org}) burial during much of the Cambro-Devonian are also suggested by light (negative) ratios of ^{13}C to ^{12}C in marine carbonates ($\delta^{13}C$; figure 7.1). Negative (light) $\delta^{13}C$ values indicate decreased productivity or erosion and oxidation of marine or terrestrial C_{org} reservoirs, whereas positive (heavy) $\delta^{13}C$ values indicate increased marine or terrestrial photosynthesis. Indeed, Berner and Raiswell (1983) calculated low rates of C_{org} burial (relative to modern values) for much of the Cambro-Devonian (see also Berner, 1989). As terrestrial forests were not well-established much before the Devonian (Knoll and Rothwell, 1981; Knoll and James, 1987; Berner 1989, 1992; Horodyski and Knauth, 1994), it is unlikely that terrestrial photosynthesis significantly influenced the $\delta^{13}C$ curve much before this time (Raiswell and Berner, 1986; Berner, 1989, 1992, 1993).

The hypothesis of low productivity during the Cambro-Devonian is supported by $\delta^{34}S$ values during this interval (figure 7.1). High $\delta^{34}S$ values during most of the Cambro-Devonian suggest extensive SO_4^- reduction, as in anoxic basins (decreased deep-ocean circulation rates

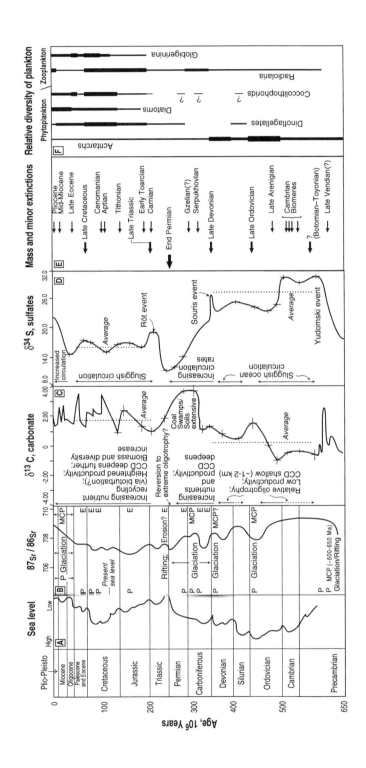

FIGURE 7.1 Sea-level, stable isotope, and lithologic indices of nutrient fluxes and productivity, extinction episodes, and major changes in plankton assemblages through the Phanerozoic. Sea level information is from Hallam (1992). $^{87}Sr/^{86}Sr$ ratios are from Holser et al. (1988); Late Precambrian portion is based on Kaufman et al. (1993). Deviation from present $^{87}Sr/^{86}Sr$ ratios and calculated runoff rates may reflect tectonics and glaciation (Raymo, 1991; Richter et al., 1992) and deforestation and agriculture (Mackenzie and Morse, 1992). P = phosphorite peak (Cook and Shergold, 1984); MCP episodes (eutrophication episodes) after Worsley et al. (1986); E = additional eutrophication discussed in text (not formally recognized as MCP episodes by Worsley et al., 1986). $\delta^{13}C$ data are primarily from Holser et al. (1988); Late Precambrian–Early Cambrian from Magaritz et al. (1986); see also Kaufman et al. (1993); Cretaceous–Recent after Renard (1986). Each datum point (center of crosshair) represents single analyses aggregated at 25×10^6-yr intervals; crosshairs indicate 1 standard error of the mean for each interval (Holser et al., 1988). Positive values indicate increased marine or terrestrial photosynthesis, whereas negative values indicate decreased productivity or oxidation of marine or terrestrial C_{org} reservoirs (release of ^{12}C-rich CO_2). General shift toward $\delta^{13}C$ positive values through the Phanerozoic suggests increased marine primary productivity and increased marine C:P burial ratios. Averages (dotted lines) fitted visually. $\delta^{34}S$ data are from Holser et al. (1988); events after Claypool et al. (1980). Each datum point (center of crosshair) represents single analyses aggregated at 25×10^6-yr intervals; crosshairs indicate 1 standard error of the mean for each interval (Holser et al., 1988). High average values suggest extensive SO_4^{2-}-reduction in widespread anoxic basins (Cambro-Devonian, Mesozoic). Gradual shift toward lower values in late Paleozoic suggests increasing deep-water overturn rates and oxygenation. Pronounced excursions to high values (events) suggest mixing into the photic zone of nutrient-rich anoxic waters previously stored in isolated basins (Claypool et al., 1980). See text for discussion of Neogene $\delta^{34}S$ excursion. Averages fitted visually. In mass and minor extinction episodes of the Phanerozoic, thickness of arrows indicates intensity (modified from Sepkoski, 1992; cf. Sepkoski, 1986; late Early Cambrian extinction based on Signor 1992, 1994). Geologic range and relative diversity of selected plankton groups are from Martin (1995a). Shift from phytoplankton characteristic of presumed superoligotrophic conditions during the Cambro-Devonian (acritarchs) to taxa characteristic of mesotrophic conditions in the Mesozoic (dinoflagellates) to diatoms (which prefer nutrient-rich or eutrophic waters) in the Neogene suggests secular rise in nutrient availability and productivity through the Phanerozoic (geologic ranges and diversity based on references in Martin, 1995a; see text for further discussion). *Modified from Martin, 1996b; placement of Precambrian–Cambrian boundary after Kaufman et al., 1993.*

and oxygenation). High $\delta^{34}S$ values during the Cambro-Devonian presumably resulted from oxidation of available C_{org} by oxyphobic SO_4^--reducing bacteria (using $SO_4^=$ as an electron acceptor; see equation 4.6). Oxidation of marine C_{org} by sulfate-reducing bacteria in turn may have contributed to low $\delta^{13}C$ values (by releasing ^{12}C-rich CO_2) and to extensive anoxia (Berry and Wilde, 1978; Berner and Raiswell, 1983; Raiswell and Berner, 1986; Holser et al., 1988). Moreover, high atmospheric CO_2 levels (Berner, 1993) may have enhanced preservation of C_{org} despite low rates of marine photosynthesis (Garrels et al., 1976; Benton, 1979). The overall parallelism between the $\delta^{34}S$ curve and the cerium curve (which has been suggested to indicate reducing conditions) corroborates increased anoxia of bottom waters during the Early to Mid-Paleozoic (Wright et al., 1987).

The overall low productivity conditions during the Cambro-Devonian were punctuated by short intervals of heightened nutrient availability. Positive $\delta^{13}C$ shifts in the Late Precambrian (e.g., Magaritz et al., 1986; see also Kaufman et al., 1993; Brasier et al., 1994; Knoll, 1996), apparently in conjunction with glaciation (Kaufman et al., 1993), suggest increased nutrient availability during this time. A pronounced excursion of $\delta^{34}S$ to high values (Yudomski event, figure 7.1) also suggests mixing into the photic zone of anoxic (and perhaps nutrient-rich) waters that may have been stored in isolated basins or deep in the oceans (Claypool et al., 1980; Keith, 1982; Wilde and Berry, 1984; Vogt, 1989; see also Kaufman et al., 1993). The broad Late Cambrian $\delta^{13}C$ shift occurred in conjunction with a major transgression, which may have also stimulated marine photosynthesis by releasing nutrients trapped in shelfal sediments through erosion and oxidation of sedimentary organic carbon (Broecker, 1982; Compton et al., 1993).

Positive shifts of the $\delta^{13}C$ curve in the Late Ordovician and Late Devonian appear to be associated with positive excursions in the $\delta^{18}O$ curve, glaciation, and sea-level fall (Orth et al., 1986; Popp et al., 1986; Stanley, 1988; Brenchley, 1989; Brenchley et al., 1994). Another positive $\delta^{34}S$ shift (Souris event) occurred in the Late Devonian (Claypool et al., 1980). The small Late Devonian $\delta^{13}C$ shift (as opposed to the strong Late Ordovician rise) in the curve of Holser et al. (1988; figure 7.1) may have resulted from erosion of terrestrial C_{org} reservoirs during sea-level fall (Algeo et al., 1995) or poorly developed glaciers (Frakes et al., 1992) and slower deep-ocean turnover rates than in the Late Ordovician. On the other hand, Joachimski and Buggisch (1993) found positive excursions $\delta^{13}C$ of up to about 2‰ just above well-documented carbon-rich shales known as the Kellwasser horizons (see also Caplan et al., 1996). Indeed, soils were also well-developed by the Late Devonian (Knoll and James, 1987), which suggests increased weathering rates and nutrient flux from the continents by this time (Berner, 1989, 1992, 1993). In contrast, Tappan (1986) argued that increasing terrestrial biomass (plant litter, soils, and so on) would have sequestered nutrients on land and kept them from the oceans, thereby starving marine plankton and causing the collapse of the marine biosphere.

Lithologic evidence supports the hypothesis of episodic eutrophication during the Cambro-Devonian. Phosphorus is a limiting nutrient on geologic timescales and is intimately involved in energy storage and transfer and other cellular pathways (Fox, 1988), so its availability probably helped to constrain marine photosynthesis. Worsley et al. (1986; see also Sheldon, 1980) recognized a series of steplike increases in marine carbon-to-phosphorus (MCP) burial ratios that presumably reflect increased phosphorus availability and permanently enhanced marine productivity (C_{org} burial rates) during the Phanerozoic (figure 7.1). According to this scenario, extensive phosphorite deposition during MCP episodes resulted from intensified recycling of phosphorus back to the photic zone as a result of glaciation, better oxygenation of shallow waters, and enhanced rates of bioturbation (phosphorus scavenging; Worsley et al., 1986).

According to Worsley et al. (1986), the first MCP episode of the Phanerozoic began in the late Precambrian (about 600–650 Ma; figure 7.1; see also Brasier, 1992; Zhuravlev and Wood, 1996). This episode appears to be associated with glaciation, mixing of nutrient-rich (presumably anoxic) waters into the photic zone (Yudomski $\delta^{34}S$ event), and the pronounced positive excursions in $\delta^{13}C$ described previously (figure 7.1). A second MCP episode (again in association with a positive $\delta^{13}C$ excursion) occurred in the Late Ordovician. Worsley et al. (1986) do not indicate an MCP episode in the Late Devonian, but there is a small rise in the frequency of phosphorites that approximately corresponds to the carbon and sulfur isotope excursions noted previously (Cook and Shergold, 1984).

Microfossil and other biotic indices support these interpretations. Low overall phytoplankton densities during the Cambro-Devonian are suggested (counterintuitively) by the record of acritarch diversity (Martin, 1995a, 1996a). Acritarchs were the dominant phytoplankton of Cambro-Devonian seas (Tappan, 1980) and, although of uncertain affinities, are normally considered to be cysts of marine eukaryotic unicellular algae that were resistant to inimical conditions (Tappan,

1980). Many workers consider high acritarch diversity in ancient rocks to indicate nutrient-rich conditions and high productivity (see Tappan, 1968, 1970, 1971, 1980, 1982, 1986; Tappan and Loeblich, 1971, 1973; Bambach, 1993:390–391), but modern plankton diversity is lowest in nutrient-rich regimes, and, conversely, highest in oligotrophic waters (see Margalef, 1968, 1971; McGowan, 1971; Hallock, 1987; Siesser, 1993). Like modern plankton, acritarch assemblages exhibited decreasing diversity in a basin-to-shelf (presumably oligotrophic-to-eutrophic) direction (Tappan, 1980); thus high cyst diversity in the Cambro-Devonian (figure 7.1) may reflect primarily superoligotrophic conditions rather than nutrient-rich conditions.

Fossil zooplankton also suggest adaptation to overall low phytoplankton densities during much of the Cambro-Devonian. Graptolites appear to have lived in low-oxygen, nutrient-rich waters just below the photic zone, and may have migrated upward to feed on occasional phytoplankton blooms caused by intrusions of deeper nutrient-rich waters into the photic zone (Berry et al., 1987). Conodonts may have lived similarly, as they also exhibited depth-stratification (Clark, 1987). Radiolarians during this time apparently lived either in highly productive shallow waters, in oligotrophic oceanic gyres with symbiotic algae, or in deeper subphotic layers of the open ocean as detritivores and bacterivores (Casey, 1993).

Studies of fossil macrobenthos also suggest that food availability was low during much of the Cambro-Devonian. The Cambrian fauna was dominated by trilobites (figure 5.1), which were vagile (moving) deposit feeders (ingesting bottom sediment for organic matter and associated microbiota). The trilobite-rich fauna began to give way to a brachiopod-dominated fauna in the Ordovician. Based on experiments with modern brachiopods, which today characterize oligotrophic habitats, Paleozoic brachiopods had very low food (energy) requirements and were able to survive intervals of starvation of 2 or more years (Rhodes and Thayer, 1991). Other relict Paleozoic taxa (living fossils such as the coelacanth and *Nautilus*) appear to have low metabolic rates and survive today in oligotrophic refugia such as caves or deep waters (Vermeij, 1987, 1994; Thayer, 1992; Boutilier et al., 1996; cf. Knoll et al., 1996); food supplies are presumably too low in such refugia to sustain the high metabolic levels of competitors and predators that arose much later (Thayer, 1992; Bambach, 1993).

Intervals of low acritarch diversity during the Cambrian and Silurian presumably reflect the elevated nutrient levels that followed the Late Precambrian and Late Ordovician eutrophication episodes, respectively (that is, abundant nutrients precluded cyst formation; see also Vidal and Knoll, 1982; Knoll and Swett, 1987). Through the Cambrian, following the Late Precambrian MCP episode, levels of dissolved nutrients presumably declined as they were progressively incorporated into plankton (and other) biomass along food chains. Diversification of epifaunal (above-bottom) suspension-feeders first took place during this time (Ausich and Bottjer, 1991) as eukaryotic plankton presumably became abundant in the waters above bottom (Bambach, 1993; Signor and Vermeij, 1994; Martin, 1995a, 1996a); this occurred in part through the process of tiering, in which organisms feed at different depths above bottom and so more finely subdivide the available habitat and niche space (Ausich and Bottjer, 1991). As nutrients were progressively sequestered into biomass, cyst diversity rose to a peak in the Ordovician (that is, there was a return to low levels of *dissolved* nutrients; figure 7.1). A similar decrease and increase in acritarch diversity followed the Late Ordovician MCP episode in the Silurian and Devonian, respectively (figure 7.1). Interestingly, the depth and intensity of bioturbation increased during the Early Cambrian, again during the Middle–Late Ordovician (after sea-level transgression and presumed nutrient release), and between the Ordovician and Early Devonian (following the Late Ordovician MCP episode; Larson and Rhoads, 1983; Bottjer and Droser, 1994), which suggests greater C_{org} (food) concentrations in sediment in response to an increased "rain" of dead organic matter to the bottom.

ALTERNATIVE INTERPRETATIONS

This interpretation of carbon and sulfur isotope data as reflecting generally low nutrient levels in Cambro-Devonian surface waters is by no means unequivocal. Other workers have suggested that the large amounts of petroleum and black shales deposited during the Cambro-Devonian (see chapter 6) were instead the result of high rates of marine photosynthesis stimulated by turnover and ventilation of the oceans via the production of dense, salty water (figure 7.2; van Andel, 1994:207–218). Low C_{org} burial rates and negative $\delta^{13}C$ values during much of the Cambro-Devonian might, then, reflect the lack of well-developed terrestrial floras (less storage of ^{12}C in plant debris) before the Devonian rather than low rates of marine photosynthesis (Berner and Raiswell, 1983; Raiswell and Berner, 1986; Berner, 1989). According to this alternative scenario, warm, saline deep waters would have been unable to hold sufficient oxygen to cause unabated oxida-

FIGURE 7.2 Ocean circulation and marine photosynthesis for icehouse (oligotaxic) and greenhouse (polytaxic) states. See also table 6.2. *From T. H. van Andel,* New Views on an Old Planet. © *Cambridge University Press 1985, 1994. Reprinted with the permission of Cambridge University Press.*

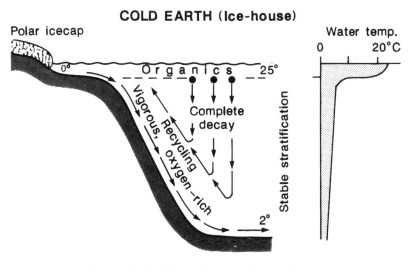

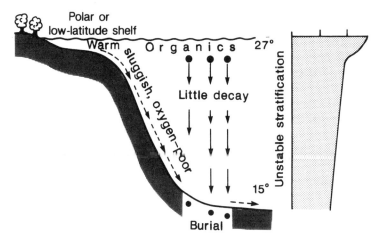

tion and decay of organic carbon; eventually, the oxygen supply in sea-water would have had to have run out. When this happened, fewer nutrients would have been released from organic carbon back to the photic zone, and marine productivity and organic carbon formation (through photosynthesis) would have slowed. With less oxygen demand (less organic carbon decay), the oxygen content of deep waters would have increased, and organic carbon decay and nutrient release would have eventually resumed, thereby stimulating productivity. According to this interpretation, carbon-rich black shales should be laminated, with very dark (carbon-rich) layers alternating with lighter layers, and this is indeed the case for Mesozoic black shales (van Andel, 1994:209).

However, Early Paleozoic (especially Cambro-Silurian) black shales appear to differ from their Mesozoic counterparts. This is not to deny that heightened productivity may have been important in the generation of some Early Paleozoic black shales, such as those in the Silurian (Moore et al., 1993; see also previous discussion of Late Ordovician MCP), but these black shales are thicker and were deposited over much larger areas than Mesozoic black shales (Thickpenny and Leggett, 1987). This suggests that anoxia was much better developed during the first half of the Paleozoic than during later times; low oxygen content of ocean waters may have been a relict of low-oxygen conditions left over from the Precambrian (Berry and Wilde, 1978; Worsley et al., 1986, 1991; McMenamin and McMenamin, 1990:161–166) or the result of high CO_2 levels in the atmosphere (Berner, 1993).

Interestingly, Ingall et al. (1993) and Van Cappellen and Ingall (1994) have suggested that nutrient regeneration is accelerated in anoxic waters. But this could have resulted in precipitation of nutrients such as phosphates rather than enhanced recycling of nutrients to the photic zone (Beier and Hayes, 1989; see also Ruttenberg and Berner, 1993, and Kump and Mackenzie, 1996). Significant regeneration of nutrients and high marine productivity during much of the Cambro-Devonian is also not supported by the fossil record, as I have already noted (see also Bambach, 1993). In summary, both views regarding the generation of carbon-rich deposits may be correct, but it depends on which carbon-rich deposits one is talking about: early Paleozoic or Mesozoic.

THE PERMO-CARBONIFEROUS

During the Devonian, extensive forests began to occupy the land, and this trend continued into the Permo-Carboniferous with the spread of extensive coal swamps (figure 7.1). The production of lignin, a compound that confers structural rigidity in land plants, may have sequestered sufficient carbon in terrestrial deposits to cause a substantial increase in atmospheric pO_2 (Robinson, 1990; Graham et al., 1995). Conceivably, increased rates of terrestrial photosynthesis may have drawn down atmospheric CO_2, thereby contributing to southern hemisphere glaciation (figures 6.11 and 7.1), sea-level fall, and enhanced ocean turnover rates during this time. Indeed, the gradual shift toward lower $\delta^{34}S$ values during the Permo-Carboniferous (figure 7.1) suggests increased oxygenation as a result of overturn of deep-ocean waters (Berry and Wilde, 1978; Parrish, 1982). Oxidation of reduced sulfur deposits (iron sulfides such as pyrite, for example) presumably acted as a negative feedback on atmospheric oxygen levels; the earth's surface redox state is balanced by the exchange of electrons between oxidized and reduced (C_{org} and sulfide) reservoirs (Garrels et al., 1976; Kump, 1989). Interestingly, though, dragonflies with wingspans of 25–30 cm have been found in coal swamp deposits; insect flight muscle is one of the most physiologically active types of muscle known to occur in the animal kingdom, and suggests oxygen levels substantially higher than those today (Robinson, 1990). (Note also that the constraint of predatory birds did not develop until the Jurassic).

Heightened nutrient levels (submesotrophic or subintermediate conditions) in the oceans during the Permo-Carboniferous are suggested by the broad rise in $^{87}Sr/^{86}Sr$ ratios and a fourth MCP episode in the Late Pennsylvanian (figure 7.1). By the Carboniferous, then, nutrient levels had apparently reached a minimum threshold value that supported large, permanent populations of plankton (Pitrat, 1970; see also Knoll, 1989). After the Devonian, acritarchs are only a minor component of the microfossil record (figure 7.1), which also suggests permanently elevated nutrient levels (see the discussion of Tappan's conclusions in chapter 5). $CaCO_3$ pelagic rain rates to the bottom should have also increased somewhat (Boss and Wilkinson, 1991) so that incipient calcareous oozes may have begun to accumulate at MOR crests by the Late Pennsylvanian (300 Ma; Boss and Wilkinson, 1991). Decreased atmospheric CO_2 (figure 6.1) and a deepening CCD (a result of erosion of cratonic limestones exposed by falling sea-level fall; figure 7.1; Boss and Wilkinson, 1991; Mackenzie and Morse, 1992) might also have enhanced calcareous ooze preservation.

Other biotic evidence for increased nutrient levels comes from the benthos. This includes the decline of the Fusulinina (benthic foraminifera) beginning about the mid-Carboniferous (figure 6.10; Tappan and Loeblich, 1988). By analogy to modern larger reef-dwelling foraminifera, the Fusulinina were adapted to nutrient-poor (oligotrophic) waters (Hallock, 1982; Tappan and Loeblich, 1988). The Fusulinina were replaced by smaller, rapidly growing foraminifera during the Middle and Late Permian (Tappan and Loeblich, 1988; see figure 6.10), and the Fusulinina themselves tended to decrease in size and lose the shell layer known as the keriotheca after the Guadalupian (Stanley and Yang, 1994). The keriotheca was presumably an adaptation to oligotrophic conditions, as it appears to have housed algal symbionts that provided food under oligotrophic conditions (cf. Ross, 1972; Hallock, 1982). The prevalence of calcareous algae in Permo-Carboniferous mounds (James, 1983) also suggests elevated nutrient levels (Hallock, 1987) after the Devonian. Increased food supplies beginning in the Carboniferous are also suggested by higher tiering (up to 1 m above bottom) in epifaunal suspension-feeding communities (Ausich and Bottjer, 1991) and increasing depth of bioturbation (Thayer, 1983; Sepkoski et al., 1991), which is energetically costly (Rhodes and Thayer, 1991; Bambach, 1993); that is, there were reliable food supplies (plankton) high in the water column and deep in sediment (detritus), respectively (Pitrat, 1970; Valentine, 1973:453; Bambach, 1993; Martin, 1995a, 1996a). Moreover, durophagous (shell-crushing) predation increased markedly in the Late Devonian (Signor and Brett, 1984), which hints at lengthening food chains and increasing metabolic rates (Rhodes and Thayer, 1991; Thayer, 1992; Bambach, 1993; Rhodes and Thompson, 1993; see also Vermeij, 1987, 1990).

PERMIAN REVERSION TO SUPEROLIGOTROPHY?

Why, then, were there no Permo-Triassic oozes? After widespread eutrophication in the Carboniferous and Early Permian, the oceans may have gradually reverted toward an oligotrophic state by the Late Permian that was reminiscent of the Cambro-Devonian. The strong negative $\delta^{13}C$ shift in the Late Permian (figure 7.1) was interpreted by Holser et al. (1991) as indicating erosion and oxidation of shelfal C_{org} reservoirs that culminated, at the Permo-Triassic boundary, in well-developed anoxic waters. This is also suggested by cerium anomalies (Wright et al., 1987; but see Erwin, 1993:204–206, for criticisms). A lack of ice caps (Frakes et al., 1992; also see figures 6.11, 7.1), and perhaps also salinity stratification of the oceans (Holser et al., 1991;

cf. Erwin, 1993:237–239), may have lowered deep-water overturn rates, so that dissolved nutrients were sequestered below the photic zone as anoxia was developing (cf. Bramlette, 1965; Lipps, 1970:10; Erwin, 1993:156); indeed, only 4 out of 11 source rocks correspond to upwelling in the Late Permian, whereas three out of four organic-rich rocks are related to upwelling in the Early Permian (Parrish 1982, 1987).

Like the negative $\delta^{13}C$ shift during this time, $^{87}Sr/^{86}Sr$ ratios also declined. But sea-level fall (figure 7.1) and erosion of continents should, seemingly, have added nutrients to the oceans, increased $^{87}Sr/^{86}Sr$ ratios, and stimulated marine photosynthesis. There are several possible explanations for this dilemma. First, the Permian was quite arid (Parrish, 1993). Based on the low $^{87}Sr/^{86}Sr$ ratios for the Permian, Tardy et al. (1989) concluded that runoff was minimal; moreover, interior drainage may have been extensive (Wyatt, 1984; François and Walker, 1992). Second, any strontium delivered to the oceans may have been concentrated in the light isotope because of erosion of extensive basaltic lavas (2–3 million km^3; Renne et al., 1995) that had been extruded onto the continents (Siberian traps; Holser and Magaritz, 1987, 1992:3302). Third, during the Permian, gymnosperm forests were replacing the lycopsid coal swamps of the Carboniferous. Evergreen leaf litter releases nutrients slowly (Tappan, 1986; Knoll and James, 1987), and may have retarded weathering of continental rocks. Indeed, charophytes ($CaCO_3$-secreting green plants of uncertain affinity), which today live in low-nutrient fresh water, were common in the Permian (Tappan, 1986), and the Fusulinacea rediversified somewhat after their initial decline (Tappan and Loeblich, 1988).

Other changes in the marine benthos are consistent with lowered productivity. Extinction rates were greater for brachiopods (despite low food requirements) than for bivalves during the end Permian extinctions, apparently because some bivalves also relied on deposit feeding (Rhodes and Thayer, 1991). If suspended food supplies decreased during the Permian, deposit-feeding would have become increasingly advantageous (Rhodes and Thayer, 1991; see also Ausich and Bottjer, 1991, and Sheehan and Hansen, 1986; cf. Levinton, 1996). Also, during the Late Permian there may have been a decline in the height of epifaunal suspension-feeding communities (Ausich and Bottjer, 1991), which may have been augmented by biological "bulldozing" of suspension feeders by deposit (sediment) feeders that plowed through the sediment (Thayer, 1983).

THE MESO-CENOZOIC

Carbon isotope values started to rise again in the Triassic, suggesting enhanced marine primary production as a result of the diversification of plankton (figure 7.1). The strontium isotope curve exhibits a sharp rise across the Permo-Triassic boundary and closely approaches the average Phanerozoic value (figure 7.1). Despite the overall aridity of the Permo-Triassic, Tardy et al. (1989) indicate a slight rise in continental runoff across the Permo-Triassic boundary, and Holser and Magaritz (1992) suggested increased erosion at this time (see also Kramm and Wedepohl, 1991). Besides erosion, the rise in the strontium isotope curve across the Permo-Triassic boundary is perhaps consistent with upwelling in incipient seaways associated with rifting of Pangaea or overturn or release of anoxic marine waters in which terrestrially derived nutrients may have accumulated (note sharp $\delta^{34}S$ spike or Röt event; cf. Yudomski event; see also Małkowski et al., 1989). A slight rise in the strontium and carbon isotope curves also occurs in the Late Triassic (figure 7.1).

During the rest of the Mesozoic, the $\delta^{13}C$ displays a series of sharp positive excursions that correspond to intervals of heightened C_{org} burial (note $\delta^{13}C$ spikes or oceanic anoxic events, or OAEs; figure 7.1). Despite apparently sluggish ocean circulation during the Mesozoic (note $\delta^{34}S$ average; see also Bralower and Thierstein, 1984; Frakes et al., 1992), enhanced circulation (driven by salinity differences, wind, or submarine volcanism, for example) at times may have stimulated dramatic rises in marine productivity. Indeed, although Worsley et al. (1986) did not recognize any MCP episodes in the Mesozoic (apparently because glaciers were only poorly developed at best; Frakes et al., 1992), extensive phosphorites sometimes occur in the vicinity of OAEs (figure 7.1). Föllmi et al. (1993) suggested that this was due to enhanced continental weathering caused by elevated atmospheric CO_2. Nutrient cycling on shelves may also have accelerated somewhat in response to rising bioturbation rates through the Mesozoic (Thayer, 1983; Sepkoski et al., 1991; see also Vermeij, 1987). Expansion of angiosperms in the Cretaceous may have also increased nutrient fluxes to shallow seas, as angiosperm leaf litter decays more rapidly than gymnosperm litter (Knoll and James, 1987; Vermeij, 1987; cf. Tappan, 1982, 1986).

Strong positive excursions in $\delta^{13}C$ occurred again in the Cenozoic, beginning near the Eo-Oligocene boundary (Shackleton, 1986). These shifts reflect increased marine productivity as a result of formation of polar ice caps, enhanced deep-water overturn, and enhanced continental erosion (through sea-level fall) and nutrient input from land (note rise in $^{87}Sr/^{86}Sr$ curve; figure 7.1). Also beginning about this time was another series of MCP boosts (figure 7.1). Indeed, Hallock et al. (1991) and Brasier (1995) have proposed extensive turnover in symbiont-bearing foraminifera during this time because of nutrient influx. The $\delta^{34}S$ shift to higher values in the Neogene (figure 7.1), despite increased deep-water overturn and presumed oxygenation, suggests that increased marine photosynthesis (organic carbon production) combined with falling sea level and sediment influx (rapid burial) to make organic carbon increasingly available for SO_4^{2-} reduction (Canfield, 1991; compare the previous interpretation of Cambro-Devonian $\delta^{34}S$ values).

The diversification of marine plankton beginning in the Mesozoic has been attributed to sea-level rise and the resultant increase in water column stratification and habitat availability (Lipps, 1970). Among the predominant groups of Mesozoic plankton are the dinoflagellates, which are often preserved as cysts, and the modern representatives of which tend to prefer mesotrophic (intermediate) nutrient levels (Kilham and Kilham, 1980). Unlike acritarchs, then, diversification of dinoflagellate cysts during the Mesozoic may signal heightened nutrient levels (although not as high as today's; Bralower and Thierstein, 1984) that may have fueled the rise in plankton diversity as pelagic habitats reappeared. Diatoms, which today prefer nutrient-rich (eutrophic) conditions (Kilham and Kilham, 1980), diversified explosively in the Miocene (Tappan, 1980) in response to further eutrophication. Bioturbation rates continued to rise through the Cenozoic (Thayer, 1983; Sepkoski et al., 1991), presumably in response to increasing surface productivity.

The rise in productivity through the Meso-Cenozoic was accompanied by significant changes in abundance and diversity of the bivalve-rich Modern fauna (figure 5.1; see also Kidwell, 1990; Bambach, 1993; Kidwell and Brenchley, 1994). Modern bivalves require much higher food levels than do modern brachiopods (Rhodes and Thayer, 1991). The diversification of the Modern fauna began in nutrient-rich Cambro-Ordovician nearshore environments (as did the trilobite and brachiopod-dominated faunas), after which it radiated into progressively deeper habitats. All three faunas followed the same onshore–offshore pattern as the primary locus of calcareous and siliceous ooze deposition (and, presumably, food), moved away from shore (Bambach, 1993; chapter 6). The advent of abundant robust bivalve shells (as compared to brachiopods) may have also imparted a taphonomic bias to the fossil record by increasing durations of time averaging (Kidwell and Brenchley, 1994).

Rising food levels in the Mesozoic may have also permitted reef-building taxa (Scleractinia) to establish extensive mutualistic relationships with photosynthetic algae. The algae live in the tissues of the corals and receive nutrients (in the form of waste products from the corals), and through photosynthesis in turn they provide food for their animal hosts (equation 6.3). By using CO_2 available in the corals' tissues, the algae presumably accelerate calcification in corals by driving equation 7.1 to the right:

$$Ca^{2+} + HCO_3^- \rightarrow CaCO_3\downarrow + CO_2 + H_2O \qquad (7.1)$$

Wood (1993) suggested that this relationship was not established until the Mesozoic, when branching skeletons among coral taxa (Scleractinia) became common. Before this time, many reef-building taxa secreted massive skeletons (such as stromatoporoids). Enhanced nutrient availability in the Mesozoic may have sustained coral growth rates sufficiently to build the branching skeletons necessary for extensive photosymbiosis, as well as the diversification of antifouling predators and grazers, which keep the corals free of other organisms that might overgrow them (Vermeij, 1987, 1994; Wood, 1993; Martin, 1995a, 1996a).

SECULAR INCREASE IN BIOMASS AND DIVERSITY

The records of marine plankton, stable isotopes, and lithology (MCP episodes) suggest that metabolic rates (level of the individual), biomass (approximate level of the ecosystem), and diversity (level of the biosphere) were all being ratcheted upward through the Phanerozoic by a steplike rise in nutrient (food) availability (figure 7.3; see also Brooks and Wiley, 1988). Indeed, perhaps the Paleozoic, Mesozoic, and Cenozoic Eras should be called the Oligozoic, Mesozoic, and Euzoic. Other workers have suggested these secular trends based on either the stable isotope record (Benton, 1979) or the fossil record alone (Vermeij, 1987, 1994; Kidwell, 1990; Bambach, 1993; Kidwell and Brenchley, 1994).

A trend similar to the marine record appears to have occurred on land (McMenamin and McMenamin, 1994). As land plants evolved they would have been faced with obtaining water and conducting it to tissues. At the same time, dissolved nutrients would have been carried along. Ultimately a "plumbing system" formed from the plant tissues xylem and phloem, which also conferred structural rigidity to land plants (via lignin) and allowed them to grow to larger size, thus establishing trees and forests. As terrestrial plants evolved, they also became intimately involved in mutualistic and parasitic relationships with fungi and other organisms, and terrestrial biodiversity and productivity, which vastly exceed those of marine ecosystems (Tappan, 1982, 1986), rose accordingly. Rising productivity on land may have also contributed to the cooling trend of the earth during the Cenozoic (Volk, 1989b).

Others have denied that such trends in productivity and biomass have occurred (see Van Valen, 1976). Schidlowski (1991), for example,

FIGURE 7.3 Number of marine animal families through the Phanerozoic. Note general increase through time despite extinctions. *From Systematics, Ecology, and Biodiversity by Niles Eldredge. Copyright © 1992 by Columbia University Press. Reprinted with permission of the publisher.*

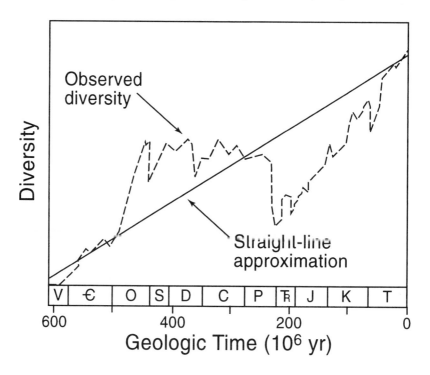

sees no trend in the carbon isotope curve or C_{org} burial rates that is indicative of rising productivity. This may be the result of looking at data over long periods of time; the variation exhibited by a population (in this case, carbon isotope values) increases the longer one counts (samples) the population (Pimm, 1991; McKinney and Frederick, 1992). Moreover, it is possible that as productivity increased through the Phanerozoic, increasing rates of bioturbation and atmospheric oxygen levels destroyed much of the carbon produced (Garrels et al., 1976), so that the fraction of organic carbon buried through time may have remained fairly constant.

Thus increased nutrient availability through the Phanerozoic could have stimulated photosynthesis through the Phanerozoic, while net C_{org} burial rates and $\delta^{13}C$ values remained fairly constant. Garrels et al. (1976) described a similar scenario involving numerical simulation of increased rates of anthropogenic erosion. In their model, they increased the global erosional rate (nutrient input to the oceans) threefold, which resulted in the tripling of organic carbon flux to the seafloor. However, a threefold increase in oxygen demand also occurred through the erosion of organic carbon and iron sulfides buried on land. Atmospheric oxygen reached a new steady-state about 15 percent below present levels in about 2 million years (even though the increased erosional rate is caused on human timescales), while atmospheric CO_2 rose to about 2.5 times current values (to about 800 ppm), which accentuates burial of carbon and sulfide in the oceans even as they are being destroyed on land. Thus the total amount of organic carbon and $\delta^{13}C$ values remained unchanged.

MORE ALTERNATIVE INTERPRETATIONS

I have suggested that nutrient input occurred repeatedly but in a catastrophic manner. Filippelli and Delaney (1992; see also Filippelli and Delaney, 1994), on the other hand, calculated that phosphorus accumulation rates in ancient phosphorite deposits and modern environments are comparable. They concluded that ancient rates fall within the range of fluxes off the modern Peru margin, although Cretaceous rates fell at the low end of the range for modern fluxes (see also Bralower and Thierstein, 1984). If modern and ancient phosphorus fluxes are indeed comparable, then nutrient fluxes to the photic zone and corresponding increases in productivity, biomass, and diversity of the marine biosphere must have occurred much more frequently through the Phanerozoic than is indicated by MCP episodes alone. In other words, MCP episodes are detectable in the geologic record because of suitable tectonic and paleoceanographic circumstances (see Compton et al. 1993; Ruttenberg and Berner, 1993), and are the most extreme versions of much more numerous phosphogenic events. If this is the case, nutrient input to the oceans may have been much more continuous than is described here.

Vermeij (1995) made many of the same points as in Martin (1995a, 1996a, 1996b), but he came to different conclusions regarding the causal agents. He hypothesized that heightened global temperature (a result of increased volcanism) promoted marine biological revolutions during the Cambro-Ordovician and Mesozoic (cf. figure 6.1). He argued that greater per capita availability of energy (ambient temperature, not nutrients) was primarily responsible for speeding up biological metabolism and resource–competitor interactions (and therefore evolution). But there is excellent evidence that nutrient (and food) availability were important in promoting biological diversification. Increased nutrient levels (through increased weathering rates, climate change, and increased rates of ocean circulation, for example) would have increased nutrient levels in the marine realm, which would have promoted increased biomass (food) levels (such as plankton) needed to have fuel evolution and expansion of the marine biosphere. Vermeij's (1995:136) denial of the importance of weathering rates during the Phanerozoic contradicts previous workers' inferences about atmospheric pCO_2 based on terrestrial plant biotas and soils, mountain building, and $CaCO_3$ saturation states, although recent studies suggest that weathering rates in the early Paleozoic and Precambrian may have indeed been higher than previously thought.

Vermeij's dialectic approach rests largely on his inferences from the record of the earth's biota alone. He does not discuss stable isotope data. Also, he argues that extinction is associated only with *decreased* primary productivity (p. 141), but isotope data also argue for prolonged intervals of *enhanced* marine productivity during some extinctions, which may have destabilized the marine biosphere (figure 7.1; see also chapter 8).

INTERACTION BETWEEN ECOLOGIC AND EVOLUTIONARY HIERARCHIES

Finally, to return to the question posed at the beginning of this chapter: How are the ecologic and evolutionary hierarchies coupled together? How is increased food (energy) translated into increased biomass and diversity? Closed systems increase in entropy (disorganization) and are the basis of the second law of thermodynamics (energy is conserved). Although they obey the second law of thermodynamics, organisms and ecosystems are *open*, dissipative systems that undergo self-organization because they exchange matter and energy with the surrounding environment, thereby increasing their internal organization. If a system is driven away from equilibrium by changing the parameters of the system, it can shift to a new stable state or attractor (see figure 5.5). For example, adding heat to a layer of fluid sandwiched between two hard surfaces at room temperature causes the fluid to self-organize into convection (Bénard) cells (just as in the earth's mantle). Changing the concentrations of reactants in a highly controlled chemical reaction, called the Belousov–Zhabotinsky reaction, causes homogeneously distributed reactants in solution to display highly organized patterns through time (Nicolis and Prigogine, 1989). In responding to the fluctuations in the physical system, the reaction's patterns change and the system is said to undergo a series of bifurcations (figure 7.4; Nicolis and Prigogine, 1989; Föllmi et al., 1993; see also May and Oster, 1976).

Similarly, if nutrient (phosphorus) supply is increased to an open system, the amount of carbon fixation (productivity) by the system, and therefore the amount of food (energy) at the base of a food web available to higher levels of herbivores and carnivores, also increases. Most of the energy transferred from one level to the next is lost as heat (about 90 percent) and becomes unavailable to make new tissue (biomass). But if more food is available, then presumably more energy can be transferred through the food web before it is lost as heat. Ulanowicz (1980) and Wicken (1980) both proposed largely theoretical mechanisms (Ulanowicz also offered an actual example) of how newly available energy sources can be converted into new links in a food web (see also Wright et al., 1993). Larger populations of different species can also be sustained because more energy is available to sustain maintenance and reproduction. Because of their greater numbers, populations are more likely to disperse over larger areas, become isolated, and give rise to new species (see also Van Valen, 1976; Vermeij, 1987:400–401).

In this way, the marine biosphere could have branched repeatedly in response to each episode of eutrophication. Nutrients cycled to the photic zone were incorporated into larger amounts of biomass (increased C:P ratios), and larger populations of organisms (Bambach, 1993). Because of their larger populations, populations (demes) were better able to exploit available niches and diversify into new taxa. Higher metabolic rates of taxa were also sustained, thereby contributing to rising diversity via predation and other biotic relationships (Bambach, 1993; Martin, 1995a, 1996a, 1996b). It may not be entirely coincidental that the diversification of rapidly growing branching corals (*Acropora*, *Porites*, and *Pocillopora*) that dominate modern undisturbed reef crests, began in the Pleistocene; similar sorts of colonies are rare or unknown from the Paleozoic (Copper, 1974; Wood, 1993). Ironically, anthropogenic eutrophication of the oceans may be accelerating a trend started by nature hundreds of millions of years ago (see also chapter 5).

These properties may have been immanent in certain species before massive eutrophication and served as a kind of preadaptation (chapter 5; see also Benton, 1995; Courtillot and Gaudemer, 1996). Opportunistic species tend to reproduce rapidly, and it is these taxa that tend to survive extinction and rediversify in the postrecovery period. In con-

FIGURE 7.4 Bifurcations presumably caused by increasing nutrient (phosphorus) levels and carbon fixation. Changes in biosphere diversity are assumed to scale upward from similar changes at ecological scales (May and Oster, 1976). *From Föllmi et al.*, Interactions of C, N, P and S Biogeochemical Cycles and Global Change, *edited by Wollast et al.* © *Springer-Verlag Berlin Heidelberg 1993. Reprinted by permission.*

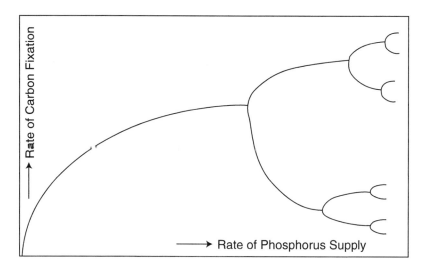

trast to stenotopic species, eurytopic (opportunistic) species have a much higher degree of "anticipation" (Salthe, 1985:147), in which they "match and therefore . . . dissipate environmental perturbations."

There are two basic ways in which entification, or the insertion of new levels (taxa) into a hierarchy, could occur: aggregation and differentiation (fragmentation) (Salthe, 1985). In aggregation, new levels accumulate by aggregation of entities at lower levels. This process has a reductionist flavor to it: Lower levels are more important than higher levels. In differentiation, instabilities cause the system to subdivide into subsystems, which, because they are semi-independent, go on to differentiate from each other so that the eventual larger system becomes more heterogeneous than previously. In fragmentation, an upper level exists from the beginning, and so it exerts control over lower levels. Adaptability turns into adaptations. "Every time a new level emerges, the entire system reorganizes to new stable points in state space. Thus aggregation occurs consequent upon differentiation" (Salthe, 1985:145). Indeed, in order to get the aggregation model working, differentiation may have to provide the units that aggregate (Salthe, 1985). In this way, complexity, the existence of entities located in the same place that do not directly interact, increases. And in complex systems, hierarchies make complexity manageable. Moreover, only complex systems with hierarchies evolve (Salthe, 1985:46; see also Valentine and May, 1996). Entification no doubt works on a variety of levels, but one of the ways it works on the grandest biological level of all—the biosphere—is, paradoxically, through another process: extinction.[2]

[2]The hypothesis of a secular rise in nutrient availability, productivity, and biomass of the marine biosphere is supported by Martin (1997, *Geological Society of America, Abstracts with Programs* 27, in press) using data of Jackson and Moore (1976, *Chemical Geology* 18:107–136).

8

Extinction

E. O. Wilson (1992), in his highly acclaimed book *The Diversity of Life,* estimated—optimistically—that the number of species doomed each year is 27,000, each day 74, and each hour 3. Wilson's estimates are sobering. Briggs (1994) calculated that at current rates of extinction, 3/4 of all terrestrial species will disappear within 200 years. We are in the midst of a mass extinction that, if it continues unabated, will surpass the greatest extinctions of the earth's history (Kauffman, 1994). Isn't it obvious? Doesn't the earth's biosphere seem to be collapsing before our very eyes? Probably not, unless one lives in a tropical rain forest or near a coral reef, the world's two most diverse ecosystems, which have been affected disastrously by human activity.

Extinction is normal, though. Raup (1991) calculated that over 99.9 percent of all the species of organisms that have ever existed have become extinct. If this were not so, speciation probably would have stopped many millions of years ago because the earth's surface would have been saturated with species. Without extinction, much fewer evolutionary opportunities would have occurred because all of the habitats and niches (ecospace) would have been filled long ago. Extinction breaks the species-rich hegemony of previously existing taxa (Jablonski, 1989); in effect, extinction produces history by altering the course of life (see Knoll, 1989, for plankton).

So why should one care about the present mass extinction? Because it is not just that species are being lost forever, but the *rate* at which they are being lost that must concern us. When species disappear as rapidly as they are now, we merely increase the chances of collapse of the global ecosystem; moreover, their replacements will evolve only slowly (on human timescales). Despite the great press that the demise of the dinosaurs has received, and its apparent cause (impact of one or more extraterrestrial bodies), there is good evidence that many of this planet's greatest extinctions were quite slow and imperceptible by human timescales (although geologically they were often fast). The Permian extinctions are estimated to have spanned between 3 and 8 million years, for example (Erwin, 1993; see also papers in Donovan, 1989b), and may actually consist of clusters of events (Hoffman, 1989a; Stanley and Yang, 1994). The causes of these extinctions are still being debated, but they, and the current extinction episode, all have one thing in common: habitat loss (Eldredge, 1991). Perhaps because extinction seems unusual (even though it is not), most of the proposed extinction mechanisms resort to unusual physical conditions to which most organisms are not adapted, and mostly ignore the interaction of the biota with their surroundings and the long-term dynamics of ecosystems, of which we know next to nothing (Helen Tappan's work is a major exception).

WHAT IS AN EXTINCTION?

We may define an extinction as an interval of time in which the rate of origination of species is exceeded by their rate of extinction, so that there is a net loss of species (see Sepkoski, 1992; Donovan, 1989a). The calculation of rates, of course, depends on two things: the accuracy of the fossil record (see chapters 3, 4, and 5) and the accuracy of the ages used to give durations of time (see chapter 2).

There are two basic types of extinction: background and mass extinction (figure 8.1). Background extinction is a continuous, low rate of extinction that is much lower in magnitude than even minor extinctions (cf. figures 8.1 and 7.1). Background extinction is punctuated by larger episodes of extinction, especially the "Big Five": Late Ordovician, Late Devonian, Late Permian, Late Triassic, and Late Cretaceous (figure 8.1). Note that these extinctions and also some of the minor ones occur near or at period–system boundaries, and help serve as the basis for the principle of faunal succession discussed in chapter 2. Despite the notoriety of the Late Cretaceous extinctions, the Late Permian extinction (the "Mother of Mass Extinctions"; Erwin, 1993) vastly exceeded the one at the end of the Cretaceous (K/T), when the dinosaurs finally disappeared from the earth. It has been calculated that about 60 percent of marine fossil genera (and well over 90 percent of the species) went extinct by the end of the Permian, whereas only about half the number of marine genera went extinct at the end of the Cretaceous.

Why use families or genera as the scale to measure extinction? Why not just species? Because some families (genera) are represented by many genera (species) and others by only a few. If one counted only

FIGURE 8.1 Rates of extinction (per million years) for families of marine invertebrates and vertebrates. Big Five mass extinctions (Late Ordovician, Late Devonian, Late Permian, Late Triassic, and Late Cretaceous) are indicated by letters A–E and occur above background extinctions (*dots*). Note general decline in background extinction rates through the Phanerozoic; this trend is statistically significant. Dashed lines indicate 95 percent confidence intervals. *From* Phanerozoic Sea-Level Changes *by Anthony Hallam. Copyright © 1992 by Columbia University Press. Reprinted with permission of the publisher; after Raup and Sepkoski, 1982.*

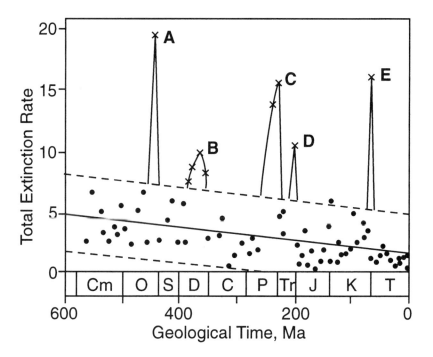

FIGURE 8.2 Histogram of extinction frequency versus intensity, and the resulting natural log–log relationship (power law) between extinction frequency and intensity, which is similar to Raup's (1991) kill curve (see text for further discussion). Note that there are fewer large extinctions than predicted by such a power law (straight line). *Modified from Kauffman, 1995.*

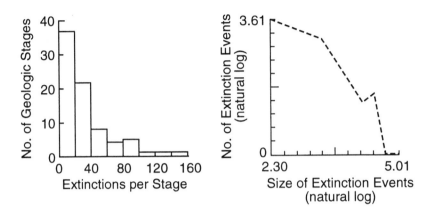

species, for example, some genera would be seriously underrepresented compared to others, and one would be comparing apples and oranges. Using genera (or, in some cases, families) weights different taxa more equally (Raup, 1991).

Mass and background extinctions appear to grade into one another. If that is the case, then what can be said about process? Raup (1991) used a histogram of the intensity of extinctions (see his figure 4–4) to calculate a kill curve (Raup, 1991, his figure 4–5), which relates the frequency of any extinction to the magnitude of its destruction (figure 8.2). This is similar to waiting for a hundred-year flood. Such a flood will occur, on average, once every 100 years, and has a much larger impact than a smaller flood that occurs on average every decade (see also chapter 5). Similarly, an extinction equal to the average size of one of the Big Five occurs about every 100 million years ("waiting or return time"). Smaller extinctions occur more often, but are less destructive. In other words, species are immune from extinction most of the time. During extinction the overall distribution of a genus is most important (Jablonski, 1986a, 1986b, 1989) because widespread species-rich genera are more likely to survive in refuges.

Selectivity of extinction may be an important clue to the causes of extinction because taxa do not all respond in the same way to strong environmental perturbations. Ammonoids, for example, underwent numerous episodes of extinction and radiation (which is why they are so useful in biostratigraphy; see chapter 2), especially in association with sea-level change and anoxia. This suggests that they were sensitive to environmental change (House, 1985, 1989). Most recently, McKinney (1995) examined extinction in echinoids at lower (genus, species) taxonomic levels. Certain traits confer vulnerability to extinction, such as low abundance, which may be associated with the fixation of deleterious mutations (Lande, 1994); large body size, which is normally associated with higher position in food chains and also low abundance; and narrow niches and associated specialization. McKinney (1995) concluded that these sorts of traits are not uniformly spread throughout the species of a genus or the taxa belonging to a higher category; rather, they are phylogenetically nested in certain taxa (see also the discussion of emergent properties of species in chapter 5). This means, according to McKinney (1995), that a dichotomy between mass and background extinction based on a *few* traits does not normally occur; instead, because extinction-biasing traits are distributed among different taxonomic levels in a nonrandom manner, extinction selectivity occurs at many taxonomic levels and at many spatio-temporal scales.

Thus, despite the apparent continuum of extinction intensity, the biotic factors involved in extinction may not necessarily scale upward from background to mass extinction. Indeed, Jablonski (1986a, 1986b, 1989) concluded that marine molluscs with planktic larvae (that are said to be planktotrophic, or feeding on other plankton) are more likely to survive during background times than lecithotrophic larvae (those that develop from an egg; Jablonski, 1986a, 1986b, 1989; Raup, 1994). Planktotrophic species tend to be widespread, so species-rich planktotrophic genera are more likely to survive an extinction than lecithotrophic species, which disperse only over small distances. During mass extinctions, however, this dichotomy breaks down, when broad geographic range of the genus itself is most important (Jablonski, 1986a, 1986b, 1989; Raup, 1994).

IMPACTS AND THE K/T EXTINCTION

During the K/T extinction, vast amounts of debris were presumably thrown into space, which blocked sunlight from reaching the earth's surface, shut down photosynthesis, and suddenly turned the earth into a gigantic refrigerator. At first glance, this is an easy scenario to test using the geologic record. An unusually high concentration of the element iridium (background concentrations are about 0 ppb; Orth, 1988), which belongs to the same group as platinum on the periodic table, was proposed to accumulate with the dust settling from the atmosphere after the impact. The iridium was presumed to have derived from the bolide itself, because this element is not normally found in rocks of the earth's crust (although it does occur in the mantle). Investigators immediately seized on this relationship (hypothesis) and searched (tested) for iridium at other K/T sites as proof of impact; ultimately, this approach was extended to extinctions of other ages, with mixed success (Claeys et al., 1992; Dao-Yi and Zheng, 1993; Kerr, 1996; McGhee, 1996). Unfortunately, iridium may concentrate in sediment under anoxic conditions, so it is not an entirely reliable indicator of impact (Holser and Magaritz, 1992; Sawlowicz, 1993). Shocked mineral assemblages (such as quartz) with *well-developed* laminations and microtektites are considered the best indicator of an impact because the pressures required to generate such mineral assemblages are presumably generated only by impacts (B. P. Glass, University of Delaware, personal communication, 1994). In fact, Alvarez et al. (1995) have traced shocked quartz into North America from the Chicxulub crater (Yucatán Peninsula, Mexico), which is now considered the most likely candidate for the K/T impact site (but see Officer and Page, 1996, for evidence against the Chicxulub site being an impact crater). Other workers have concluded that apparent stepped extinctions below the K/T boundary represent multiple impacts (Hut et al., 1987), but these may simply be taphonomic artifacts (Raup, 1989; see also chapter 2).

Raup (1991) assumed that geologically significant extinction is caused by extraterrestrial impacts. By doing so, he related an approximate crater (and impact) size to the magnitude of an extinction and generated an extinction–impact curve (figure 8.2; see also Raup, 1991, figure 10-1). If one accepts that impacts have occurred on this planet (and no doubt they have; why should this planet be any different from the rest of the solar system?), Raup's approach makes perfect sense. It is another example of the presumed upward scaling from small to large phenomena. It is also essentially a reductionist approach (coming from a paleontologist) because it attempts to simplify a system to its utmost and assumes no history (see also Hoffman, 1989a): All extinctions are caused by the same mechanism (impact), and all biological communities may be treated as being roughly identical. Only the size of the impact matters.

One problem with this hypothesis is that the relationship between impacts and extinction has not been fully tested. Certainly, good evidence (shocked mineral assemblages or microtektites) has been found for impacts occurring at the end of the Cretaceous and the end of the Eocene (a crater in the subsurface of Chesapeake Bay region about 35 million years old appears to have been confirmed; Koeberl et al., 1996; Kerr, 1995c), when there was significant biotic turnover. Others have claimed that the association of iridium with other extinctions also confirms the impact hypothesis, but as we have seen, iridium may be concentrated for other reasons (anoxia). To truly test the impact hypothesis, intervals when there was no significant biotic turnover should be examined for indications of impact, and the occurrence of craters versus extinction should be determined. The former would be much like looking for a needle in a haystack, whereas the second relationship (or lack thereof) indicates that the impact hypothesis does not fare too well (Officer and Page, 1996:33). Moreover, although a crater and microtektites do occur near the end of the Eocene, for example (see Koeberl et al., 1996), there is a great deal of evidence to suggest that any impact was overshadowed by the historical consequences of other paleoclimatic and paleoceanographic factors (Prothero, 1994).

TRADITIONAL CAUSES OF EXTINCTION

There is certainly a great deal of discussion about the mechanisms of extinction, and at least several other factors are considered to be viable candidates. Hallam (1989; 1992:180–195) gives a brief summary on which the account of traditional agents of extinction is based.

Sea Level

This is perhaps the extinction mechanism with the longest pedigree (unless one considers divine intervention, as paleontologist Georges Cuvier did early in the nineteenth century). It is not really a surprising choice, either, considering the preoccupation of most earth scientists with sea level and water depth; other important geological phenomena go hand in hand with these two factors, such as wave and current energy, lithology, and oxygen content (these latter phenomena are epiphenomena because they are what one observes, but not the cause of what is observed). As early as Newell (1967), a correlation was noted between sea-level fall and major extinctions (see figure 7.1). Newell's (1967) idea was that as sea level fell, habitat area would become more restricted and species would be driven to extinction. This would be especially true in shallow, widespread epeiric seas, in which even a minor fall in sea level would expose large areas of land. Later, Valentine and Moores (1972) related sea-level change to plate tectonics and global marine biotic diversity: As epeiric seas spread, habitat and species diversity went up accordingly; conversely, as continents collided and shelves were destroyed, habitat and biotic diversity declined (cf. figures 6.1 and 7.3).

The relationship between regression and habitat is based on the species–area curve:

$$S = cA^z, \tag{8.1}$$

where S = number of species, A = area, and c and z are constants that vary between species and habitats. This equation was originally developed for oceanic islands as part of the theory of island biogeography, which assumes that an equilibrium level of diversity is maintained by immigration and extinction (MacArthur and Wilson, 1967; see also Slobodkin, 1996). This equation was used by Wilson (1992) to make the estimates of extinction rates mentioned at the beginning of the chapter. For our purposes, c may be ignored (it is used only to change the proportional relationship, $S \alpha A^z$, to equation 8.1). The constant z, on the other hand, is very important. It is basically an estimate (not easily made), falling between 0.15 and 0.35 (for terrestrial organisms; Wilson, 1992), of the dispersal ability of a species. Species with high dispersal powers, such as birds, have low z values, whereas land snails have high z values. The higher the z value, the greater the extinction rate (Wilson, 1992:276). A number of workers have tested the relationship between sea-level fall, area, and extinction statistically and found the relationship to hold up rather well (see Hallam, 1992, for discussion). But reduction in habitat (area) involves other factors, such as increased crowding and competition, so equation 8.1 is not very helpful in determining the ultimate causes of extinction.

Global Cooling

Besides the species–area curve, there are other problems with sea level as the primary cause of extinction. Stanley (1988) noted that the major sea level fall of the Oligocene is not accompanied by mass extinction, although there is significant biotic turnover (such as development of the deep-sea psychrosphere and changes in terrestrial floral and mammalian communities; see Prothero, 1994, for extensive review); nor is the Late Silurian sea-level fall associated with extinction (figure 7.1; Hallam, 1992). This led Stanley (1988) to propose that global cooling (glaciation) is much more important in driving extinction, especially because the broad distributions of organisms seem to be determined by temperature, and tropical organisms are especially vulnerable to lower temperatures. Hallam (1989; based on Shackleton's 1986 analysis of the oxygen isotope curve), who is a major proponent of sea-level change as an agent of extinction, has countered that major climate change preceded biotic turnover (as evidenced by biostratigraphic datums) in the Eo-Oligocene, and that tropical species are also stenotopic, not just stenothermal. Moreover, by the end-Permian, the Gondwanaland (southern hemisphere) glaciers had disappeared and the earth appeared to be warming significantly (Erwin, 1993), and evidence for glaciation in the Late Devonian is somewhat equivocal (Frakes et al., 1992). On the other hand, there is good evidence for glaciers at the end of the Late Ordovician and the end of the Carboniferous (Frakes et al., 1992; figures 6.1 and 6.11), when significant extinctions occurred (figure 7.1).

Stanley (1984) also documented the regional extinction of molluscs in the western North Atlantic during the Plio-Pleistocene, when sea level was rising and falling as glaciers waxed and waned (chapter 6). Why were extinctions during the Quaternary not of greater magnitude? Perhaps sea-level lowstands were too short to effect significant extinction (Hallam, 1989, 1992; see also Pandolfi, 1996). The much slower (but often greater) regressions of the Phanerozoic lasted much longer, and occurred in shallow, epeiric seas that have no close modern analog (Hallam, 1981). Then again, a fall in sea level exposes more area of a large continent than a small one (Hallam, 1992), so that continental configuration (and climate) and overall position of sea level (figure 6.1) may constrain smaller-scale changes in sea level and climate (chapter 6).

Hallam (1992) also suggested that greater ventilation of the deeper portions of the oceans in the Quaternary and Oligocene may have provided refuges during these times. It is difficult, however, to imagine extensive migrations from the deep sea back onto the shallow shelves, at least over such a short interval as the Quaternary (cf. figure 5.1), and because the boundaries between shelf and slope conditions and their faunas are fairly distinctive, there ought to have been significant turnover in these environments if this were the case. Moreover, the migrations that do occur seem to work in reverse (shallow to deep) and are permanent. As we saw in chapter 5, living fossils tend to occur in restricted habitats such as the deep sea. On the other hand, modern shallow-water faunas could probably persist around the margins of oceanic islands (refuges) despite extensive sea level fall (Jablonski, 1985). So there are problems with both sea level and global cooling. What else could drive extinction?

Volcanism

The mechanism commonly used as a counterpoint to impact is that of volcanism. Volcanoes are fed, ultimately, by the mantle, so they could conceivably spew forth sufficient quantities of iridium, which is abundant in the mantle, to account for the layers at the K/T boundary. They also inject tremendous amounts of aerosols in the short term that drastically cool the earth, whereas the CO_2 emitted results in a *long-term* warming (chapter 6; see also Rampino et al., 1988). Although substantial volcanism occurred near the K/T boundary in India 65 million years ago (Deccan traps, where *trap* refers to the steplike appearance of the plateau that the lavas form) in association with a hotspot (mantle plume), these are flood basalts: gentle fissure eruptions that probably did not inject aerosol and iridium high enough into the atmosphere to produce climate change and a global stratigraphic signature. Moreover, the pressures generated would be insufficient to produce well-developed shocked mineral assemblages, and a number of well-documented volcanic episodes do not coincide with mass extinctions, such as the Columbia River basalts, which are flood basalts. When volcanism, extinction, and sea-level fall do coincide, as they appear to have at the end of the Permian (Erwin, 1993), it suggests that all may be controlled by the production of mantle plumes at the core–mantle boundary (see Loper et al., 1988). Mantle plume dynamics may also account for the association of end-Permian and end-Cretaceous extinctions with magnetic reversals (Loper et al., 1988; see also Larson and Olson, 1991). Rampino (1993) suggested that hotspots may be generated by impacts; moreover, the energy generated by a 10-km impactor is equivalent to a magnitude 11–12 earthquake (Rampino and Caldeira, 1993). Such impacts could produce sufficient energy to concentrate seismic energy at other foci on the earth's surface and cause crustal weakening, thereby allowing more mantle plumes to come to the surface (Rampino, 1993; see Officer and Page, 1996, for criticisms). Once again, presumed causal agents of extinction may be coupled.

Anoxia

Wilde and Berry (1984) noted that episodes of anoxia and black shale deposition appear to coincide with the Big Five mass extinctions. Hallam (1989, 1991, 1992), Wignall and Hallam (1992), and Wignall and Twitchett (1996) later extended this argument to other extinction episodes. The spread of anoxic bottom waters would be promoted by transgression, which would have been rapid (geologically speaking) in shallow, epeiric seas. The spread of anoxic waters could also have introduced massive amounts of nutrients into shallow waters and further destabilized marine ecosystems, much like the input of sewage effluent into oligotrophic lakes that causes massive blooms of algae. Erwin (1993), who reviewed a number of extinction agents with respect to the end-Permian extinction, criticized the anoxia hypothesis on a number of grounds, mainly revolving around the timing of the events. Many marine taxa, such as suspension-feeding blastoids, died out millions of years before the spread of anoxic waters.

MURDER ON THE ORIENT EXPRESS

Like Raup's (1991) approach, the mechanisms discussed earlier in this chapter are certainly plausible, but the approach in each case is reductionist and ahistorical. Extinctions are caused basically by one agent (or perhaps two), which depends on one's pet hypothesis. Hoffman (1989b), on the other hand, considers the coincidence of unusual conditions to be vitally important in producing extinction. It is somewhat like flipping a coin, except that, given the complexity of the earth's climate, there are more than two possible outcomes. The probability of heads is 1/2 for one coin flip (assuming it is a fair coin), whereas the probability of two heads in a row is 1/4, three heads is 1/8, and so on. The general formula, then, is

$$\text{Number of heads in a row} = \tfrac{1}{2}^n, \qquad (8.2)$$

where n = number of coin flips. Hoffman (1989b) maintains that extinction is very similar. The probability (coincidence) of each extinction mechanism is quite low and the probability of the coincidence of one or more occurring together even lower, hence the rarity of mass extinctions and the much greater frequency of minor ones (figure 7.1). Moreover, the analogy is a binary one (heads *or* tails), whereas in nature there would be all sorts of gradations between any extremes.

Erwin (1993) came to much the same conclusion. The end-Paleozoic extinctions were not caused by a single agent, but by a number of them acting in concert (like Agatha Christie's novel; figure 8.3), which helps to explain the severity of these extinctions. At the simplest level (largest or highest holon), this mechanism involves major, long-term sea-level regression (figure 7.1). Sea-level regression would have resulted in decreased habitat area, increased climatic seasonality (both leading to ecologic instability), erosion and oxidation of shelfal organic carbon reservoirs, and release of methane and other gases trapped in shelf sediments as the pressure of the overlying water was removed. Basalt production (Siberian traps), oxidation of organic carbon, and release of gases such as methane would have released vast amounts of CO_2 into the atmosphere (Renne et al., 1995; Kerr, 1995d), warming the earth (there is no evidence for glaciers in the latest Permian; Frakes et al., 1992) and producing extensive anoxia.

FIGURE 8.3 The "Murder on the Orient Express" scenario for the late Permian extinctions. *From* The Great Paleozoic Crisis *by Douglas H. Erwin. Copyright © 1993 by Columbia University Press. Reprinted with permission of the publisher.*

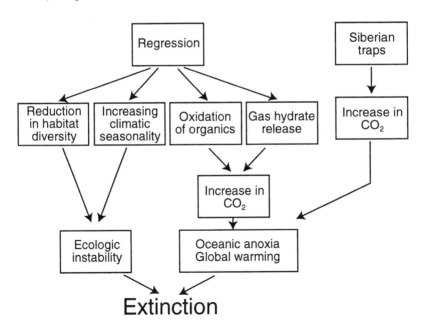

The conjunction of several of these same agents could have produced other extinctions, as Hoffman (1989a, 1989b) suggests. The configuration of the continents could serve as a low-frequency constraint acting from the highest holon. Sea level, habitat area, volcanism, and general climatic conditions would represent lower holons constrained by plate configurations and either act at higher frequencies, sometimes reinforcing one another and sometimes not, or act in different proportions from one time interval to the next. Each extinction is the result not of a unique set of causes, but of a unique *combination* of causes ultimately constrained by the overall configuration of continents and oceans. Unlike the previous approaches, this one is distinctly historical.

"CATASTROPHIC" FLUCTUATIONS IN NUTRIENT LEVELS

The decline in overall extinction intensity during the Phanerozoic (figure 8.1) has been attributed to the accumulation of geographically widespread (eurytopic), species-rich clades through time (see Flessa and Jablonski, 1985; Vermeij, 1987:400–401). As discussed in chapter 7, the stable isotope, lithologic, and fossil records suggest that marine biomass and diversity were also being ratcheted upward through the Phanerozoic by a steplike rise in nutrient (food) availability. If nutrient availability does exert strong control over the diversity of the biosphere in the long-term (Phanerozoic), could it have played a role in extinction? Such effects would presumably scale upward from ecological spatio-temporal scales (see DeAngelis, 1992; see also chapter 7).

Catastrophic fluctuations in nutrient levels (as suggested by $^{87}Sr^{86}Sr$ curves and phosphogenic episodes) described in chapter 7 appear to coincide with many of the mass (and minor) extinctions described by Sepkoski (1986, 1992; figure 7.1), and may have exacerbated extinctions via shortening of pelagic food chains (Martin, 1995a, 1996a, 1996b). In most cases these fluctuations involved substantial nutrient inputs to the surface oceans caused by enhanced chemical weathering, global cooling, sea-level fall, and increased rates of deep-ocean overturn, and were accompanied by positive shifts in the $\delta^{13}C$ curve, suggesting heightened marine productivity. Among the Big Five extinctions, this scenario includes the Late Ordovician and Late Devonian, and perhaps also the Late Precambrian–Early Cambrian extinctions (cf. Brasier et al., 1994; Zhuravlev and Wood, 1996; see chapter 7). Heightened productivity may also have occurred during the Late Triassic mass extinction, which has been ascribed to both anoxia (see Hallam, 1992) and impact (McRoberts and Newton, 1995), because a slight rise in the strontium and carbon isotope curves occurs at this time (figure 7.1). Minor extinctions could also have been affected by heightened productivity, but appear to have involved either glaciation and sea-level fall (end-Carboniferous and Late Eocene–Oligocene, for example) or warming and sea-level rise through enhanced chemical weathering and release of nutrients from shelf sediments during transgression (such as Cambrian biomeres, Mesozoic oceanic anoxic events; see also Föllmi et al., 1993; Compton et al., 1993).

Conversely, nutrient starvation may have occurred near the end of the Permian. A lack of ice caps (Frakes et al., 1992; figure 6.11), and perhaps also salinity stratification of the oceans (Holser et al., 1991), may have lowered deep-water overturn rates, so that dissolved nutrients were sequestered below the photic zone as anoxia was developing. Nutrient starvation would have essentially "pulled the rug out from under" a marine biosphere at the end of the Paleozoic that had become accustomed to greater nutrient and food availability (submesotrophic conditions) and higher metabolic rates.

Małkowski et al. (1989) also ascribe the end-Permian crisis to a gradual but ultimately drastic decrease in nutrient availability caused by increased ocean circulation. They suggest that there was a geologically rapid switch from one paleoceanographic mode to another (cf. the section on CO_2 cycles in chapter 6). Before the end-Permian extinctions, the oceans were in an "overfed mode," in which large amounts of carbon and nutrients accumulated in stagnant waters; nevertheless, extensive organic decay (via bacterial decomposition) deep in the oceans released nutrients back to the photic zone, thereby compensating for sluggish circulation, as evidenced, according to these authors, by a strong positive excursion in the carbon isotope curve before the end of the Permian. As high levels of oxygen in the atmosphere accumulated (a result of carbon storage in the oceans), the earth began to shift toward an icehouse mode, which induced more vigorous oceanic circulation. As the earth entered this mode, it eventually switched to a "hungry mode," and the carbon stored in the oceans was brought to the surface and oxidized, releasing vast amounts of CO_2 into the atmosphere (cf. volcanism scenario). According to them, recycling of nutrients to the photic zone declined because nutrients such as phosphorus form insoluble compounds in the presence of oxygen, and the carbon isotope curve gradually shifted toward negative values because surface productivity decreased. However, the interpretations of Małkowski et al. (1989) are largely dependent on a single section in Greenland, which does not exhibit the same $\delta^{13}C$ characteristics as those described by Holser et al. (1991) from Austria and may be stratigraphically incomplete or diagenetically altered (Mii et al., 1997) (cf. chapter 2). The Austrian section exhibits a long, gradual negative shift in $\delta^{13}C$ before the final pronounced negative carbon shift occurs (figure 7.1; Holser et al., 1991), which suggests that the Austrian section is intact (Hallam, 1992).

Based on Małkowski et al. (1989), Knoll et al. (1996) suggested that end-Permian extinctions were caused by the release of massive quantities of CO_2 stored in the deep ocean during oceanic turnover. Elevated CO_2 levels would have adversely affected the metabolism of much of the earth's biota. They contend that the oceans were only weakly stratified at this time and easily subject to turnover (cf. Wilde and Berry, 1984), and that CO_2 buildup in the ocean resulted from primary production in the surface layer (overfed mode). Despite sluggish ocean circulation rates, they postulate that the release of phosphorus from decaying organic matter in deep anoxic waters would have been sufficient to further stimulate photosynthesis (Ingall et al., 1993; Van Cappellen and Ingall, 1994), thereby leading to further organic decay or positive feedback (but see the section in alternative interpretations in chapter 7). Thus Knoll et al. (1996) contend that their scenario is not consistent with nutrient collapse. It is possible, however, that if ocean circulation had been sufficiently slow in the Late Permian, phytoplankton could have largely stripped the surface mixed layer of nutrients (Herbert and Sarmiento, 1991) so that nutrient collapse would have occurred.

Based on the carbon isotope analyses of the Austrian section, Magaritz et al. (1992) proposed that two processes, each with different temporal scales, were acting at the end of the Permian. As discussed in the section on Permian reversion in chapter 7, the gradual negative $\delta^{13}C$ shift in the Late Permian (figure 7.1) was interpreted by these authors as resulting from erosion and oxidation of shelfal organic carbon reservoirs (Holser et al., 1991; Holser and Magaritz, 1992); this process acted on a timescale of 10^6 years or more. They contend that the sharp negative shift in the carbon isotope curve at the Permo-Triassic boundary resulted from a decline in surface productivity, which acted over a short time interval (cf. Kump, 1991). Interestingly, the strong negative isotope shift is associated with pyrite (anoxia) and iridium (Holser et al., 1991; Holser and Magaritz, 1992; see the section on impacts and the K/T extinction in this chapter).

CATASTROPHE AND EXTINCTION

Usually, the end-Cretaceous extinctions are said to be accompanied by a very drastic decrease in marine photosynthesis that is indicated by a strong negative $\delta^{13}C$ shift ("Strangelove Ocean"; Hsü, 1986; Zachos et al., 1989; Holser and Magaritz, 1992; see also Bramlette, 1965; Tappan, 1968). As mentioned previously, however, a number of sections containing the K/T boundary, which were once thought to be continuous, may actually be incomplete based on graphic correlation (see chapter 2; MacLeod, 1995). Wang et al. (1994) found an abrupt $\delta^{13}C$ shift in the Late Permian of British Columbia like that claimed for the end-Cretaceous, but the gradual negative $\delta^{13}C$ shift seen in the Austrian section (Holser et al., 1991) appears to be missing. Other evidence suggests that a much more gradual shift may have been occurring in the latest Cretaceous. Based on a positive shift in $\delta^{13}C$ during the early–late Maastrichtian Stage transition in Ocean Drilling Program cores from the southern hemisphere ocean, Barrera (1994) concluded that marine productivity was increasing in response to global cooling and enhanced oceanic overturn. Sea level was also falling and $^{87}Sr/^{86}Sr$ ratios (and presumably nutrient input) increasing during this time (figure 7.1). Moreover, angiosperms were expanding in the Cretaceous, which may have increased nutrient fluxes to shallow seaways (Tappan, 1986; Knoll and James, 1987; Vermeij, 1987).

Say we were to transport ourselves back to the latest Cretaceous (before the impact) or latest Permian to observe climate change and extinction. We would have to wait quite a while because with few exceptions, such as the loss of individual species, one cannot *observe* extinction (Pimm, 1991). Our perception would be that from one year to the next, from century to century, and perhaps even millennium to millennium, nothing was happening. We would be in a situation similar to today's, but the changes would be more prolonged. Human activities have accelerated rates of change compared to what we normally calculate from the geologic record, yet nothing seems to be happening.

But gradual, imperceptible change can have catastrophic results. The collapse of the former Soviet Union did not appear imminent, and then the Berlin Wall was suddenly being torn apart by people who, a few days earlier, probably would have been shot on sight. According to catastrophe theory, the Soviet Union was actually moving slowly, imperceptibly over a surface from one sociopolitical state to another until it reached a fold in the surface. At this point (called a cusp), the Soviet empire jumped from one state to another, and it was this jump

201

that we observed in the news media (see chapter 2 of Casti, 1994, for a general discussion of catastrophe theory; see Crowley and North, 1988, for its application to extinction).

We face a similar situation with regard to the planet's biological communities and biodiversity. At the time of this writing, debate has begun on renewal of the Endangered Species Act, which has saved such creatures as the spotted owl in the Pacific Northwest and aroused the enmity of loggers. Should such species be saved? If so, why? In the September 1 (1995) issue of *Science*, Eisner et al. discuss "Building a Scientifically Sound Policy for Protecting Endangered Species," and in the same issue, de Ruiter et al. report that in natural food webs (not modeled ones!) uncommon species may play a more important role in maintaining community stability than more common ones (see also Polis and Strong, 1996). Pimm (1991) came to the same basic conclusion in a review of a number of different studies. Such species are often called keystone species (Pimm, 1991). If we must choose between the species to be saved and those that we let slide into oblivion, which ones do we choose? How? Except for saving whole ecosystems (and thereby potentially angering many property owners), no one has the answer.

EXTINCTION AND LIFE HISTORY

Ironically, anthropogenic disturbance may provide some clues to interpreting extinction in the fossil record. We may be able to scale upward from anthropogenic disturbance to mass extinction in the case of the decimation of modern reef communities (bear in mind the discussion of background versus mass extinction earlier in this chapter). In this upward-scaling scenario, the response of the reef biota is assumed to be roughly the same everywhere, although the actual processes introducing nutrients to the reefs and causing their degradation change from local to regional to global spatial scales, and from ecologic to geologic temporal scales.

Many modern reefs are being destroyed by excessive sedimentation and nutrients from sewage and terrestrial runoff caused by deforestation (local extinction; see also chapter 5). Most reef-building corals prefer nutrient-poor (oligotrophic) waters as opposed to nutrient-rich (eutrophic) conditions. The reason is that oligotrophic waters are clear and allow sunlight to penetrate to the bottom in shallow water, thereby allowing the algal symbionts that inhabit many corals' tissues to photosynthesize. The corals are said to be, in general, "K-selected," which tends to promote specialization and high diversity, whereas "r-selection" tends to promote generalization and low diversity, although some coral species are more K-selected than others (Margalef, 1968; Valentine, 1971; see also table 5.2 and equation 5.3).

Elevated nutrient levels and pelagic productivity may also act on corals indirectly by affecting their reproductive (life history) strategies. Under eutrophic conditions, plankton populations increase dramatically and not only does the water become murkier, but new, rapidly breeding, opportunistic (r-selected) organisms, such as boring sponges, which feed on the new plankton populations, move in and outcompete the slower-growing (K-selected) corals. Hallam (1978) proposed a similar relationship between sea level, niche specialization, and reproductive traits for Jurassic ammonites (see also "perched" faunas of Johnson, 1974).

Corals that survive on polluted reefs appear to do so because they are less susceptible to fouling and because they reproduce by brooding (planulating) larvae to an advanced stage before release (see Tomascik and Sander, 1985, 1987a, 1987b). Such a reproductive strategy appears to correlate with small, and perhaps more opportunistic, coral species capable of survival in stressed environments (Tomascik and Sander, 1985, 1987a, 1987b). Although larvae of brooding coral species can be carried long distances, based on biogeographic range data, brooding increases local recruitment rates (and survival) by providing larvae suitable for settlement immediately after release, especially in stressful environments (the recruitment hypothesis of Edinger and Risk, 1995).

Perhaps because of differences in nutrient levels, the same genera of corals that brood in the Caribbean broadcast gametes (sperm and egg) into the sea in the oligotrophic waters of the Indo-West Pacific, as do most other hermatypic coral genera there. In the large, branching, reef-building genus *Acropora*, for example, newly broadcast spat often represent 50–80 percent of juveniles in the Indo-West Pacific, but in the Caribbean, *Acropora* is uncommon among juveniles, and typically reproduces by asexual fragmentation and recementation of branches. The situation for corals is just the opposite of molluscs, in which broadcasters are favored over brooders (Edinger and Risk, 1995).

Asexual reproduction is prominent in other coral populations living near their physiological limits, and some coral populations have been reported to be sterile at their geographic limits, such as in the eastern equatorial Pacific (Scott et al., 1988). Chronic wind-driven upwelling makes the eastern equatorial Pacific among the highest-productivity regions of the world today (Martin, 1996b). Reproduction by corals in this region is dominated by asexual fragmentation, whereas settlement of sexual larvae is severely restricted because the region is isolated from sexually reproducing populations in the Indo-West Pacific by the vast expanse of ocean and from the Caribbean by the Isthmus of Panama (Richmond, 1990; Richmond and Hunter, 1990). Remote, apparently recruitment-limited populations such as those of the eastern Pacific are susceptible to extinction (see Martin, 1996b). Before about 3 million years ago, the Caribbean and Pacific were part of a larger (Tethyan) seaway, and before that (much of the Mesozoic and Paleogene) these waters were considered part of a western extension of the circumequatorial Tethys (see chapter 6).

These sorts of changes appear to have occurred on much larger spatio-temporal scales. Based on the record of reef biota and phosphorites, Edinger and Risk (1994, 1995) suggested that global cooling in the Oligo-Miocene (see chapter 6) and massive nutrient input to the largely landlocked Caribbean Sea may have caused the demise of scleractinian reefs during the Oligo-Miocene in this region (minor extinction). Edinger and Risk (1995) concluded that brooding coral genera preferentially survived the extinction, not because of greater recruitment success but because of the ecological correlates of eurytopy (species sorting; see chapter 5 and earlier in this chapter). Reef diversity declined again following Antarctic ice sheet formation in the Middle Miocene (Kauffman and Fagerstrom, 1993). Late Ordovician and Late Devonian stromatoporoid reefs appear to have suffered a similar fate but on a much larger spatial scale (mass extinction). The Late Ordovician, and perhaps also the Late Devonian one, correspond to enhanced oceanic overturn following prolonged phases (millions of years) of warm, equable climates and high sea level (oligotrophic, warm shallow seaways; chapters 6 and 7). The demise of reef platforms could also have occurred during sea-level rise (platform drowning) and release of nutrients from reworking of shelf sediments or the transgression of anoxic, nutrient-rich waters, as suggested for the Cretaceous (Hallock and Schlager, 1986).

No matter which episode is considered, the basic result is always the same (reef degradation), but the actual process causing the nutrient input varies according to the spatio-temporal scale being considered: local extinction (sewage), regional extinction (or nearly so, in the eastern equatorial Pacific) because of upwelling and lack of recruitment, minor extinction (Oligo–Miocene of the Caribbean caused by glaciation and sea-level fall or demise of Cretaceous platforms during sea-level rise), and mass extinction (southern hemisphere glaciation, global cooling, oceanic turnover, and continental weathering adding nutrients to the oceans over much larger and longer scales).

This apparent continuum from local to mass extinction can be described in terms of attractors (chapter 5). Environmental limits and biotic potential place upper and lower boundaries, respectively, on a constraint envelope (analogous to a niche hypervolume) in which complex ecological systems (landscapes) operate over broad spatio-temporal scales (O'Neill, 1988; O'Neill et al., 1986, 1989). In the present case, physical factors such as temperature are assumed to form the upper environmental limits (analogous to the concept of limiting factors) on the constraint envelope, while feeding and reproductive strategies and environmental tolerance determine the lower limit. The system (in this case a biological community) that exists within the constraint envelope is metastable (O'Neill, 1988; O'Neill et al., 1986, 1989). That is, it moves along a pathway (manifold) to new community states (local attractors) as some limiting factor—say, nutrient levels (sewage input)—acting within the constraint envelope changes (cf. Knowlton, 1992).

But what if the upper environmental limits, such as temperature, change? Global cooling and increased rates of thermohaline-driven deep-ocean overturn, for example, would increase nutrient levels in the photic zone. Then the position of the constraint envelope itself shifts as well. Because they occur at a higher hierarchical level, such physical limiting factors change on slower (geological) timescales than the biological ones (Allen and Starr, 1982; Salthe, 1985; O'Neill, 1988). The constraint envelope itself (and its contained community states) shift along a manifold that consists of a series of local attractors toward which the faster dynamics of the system (measured variables such as productivity and biodiversity) recover (O'Neill et al., 1989). If the trajectory should move toward an unstable region (bifurcation; cf. figure 7.4), the rate of return of the faster biological dynamics to the manifold should decrease as the point of instability is approached (increased return times to recovery or decreased resilience), and there should be increased correlation of demographic stochasticity (birth, death, reproductive rates, population densities, and so on) and environmental conditions (nutrients, temperature) among widely distributed local and regional populations (Johnson et al., 1995). In the case of the eastern equatorial Pacific, for example, the manifold may have shifted as a result of the closure of the Isthmus of Panama 3.0–3.5 million years ago and the establishment of the present upwelling regime there. Since that time, reefs in the eastern Pacific have been much more temporally and spatially discontinuous as compared to the western Atlantic and Indo-West Pacific, and less likely to be preserved in the fossil record (cf. Jackson, 1992). Climate-related changes in deep-ocean overturn rates could have moved the manifold toward new attractors on much larger temporal and spatial scales (Barber, 1992), including the largest of the mass extinctions. Such changes are no doubt also accompanied by global temperature change and biogeographic restriction, leaving few refugia (because most of the globe was presumably affected) to serve as sources of larvae (Johnson et al., 1995).

In this way, nutrient increase could have affected the overall diversity of the Caribbean versus the Indo-West Pacific (Martin, 1996b). Indeed, there is a great difference in modern coral diversity between the Indo-West Pacific and Caribbean regions. Approximately 37 percent of young (Late Miocene–Recent) coral genera are restricted to the Indo-West Pacific and they are much more species-rich than Caribbean genera. These differences are most easily ascribed to the very high geographic complexity of the Indo-West Pacific region ("vicariance"). Thus the availability of numerous local (and genetically variable) sources of larvae and other propagules has no doubt promoted high diversity in this region on geological timescales.

But assuming that the presumed effects of nutrient-related perturbations on reproductive patterns scale upward, the history of nutrient-related disturbance may have also played a role in generating modern regional coral diversity patterns. With the closure of the Mediterranean (a part of the Tethyan seaway) and the rise of the Isthmus of Panama, the western Atlantic and eastern Pacific were not only cut off from significant sources of larvae or rafted individuals, but also presumably exposed to chronically higher nutrient levels, which could have affected life history traits and rates of speciation. In the Caribbean and eastern Pacific, an increased incidence of both brooding and asexual reproduction in response to elevated nutrient levels beginning in the Neogene may have decreased the chances of coral speciation in these regions (see also Johnson et al., 1995). Increased brooding may have decreased speciation rates in corals because brooded larvae settle rapidly, thereby promoting philopatry and inbreeding (Richmond, 1989). Thus brooding and asexual fragmentation in corals may decrease rates of speciation (Richmond, 1989), but when coupled with rapid growth and large colony size (*Acropora*, for example), they may also decrease the chances of extinction of surviving coral genera.

PERIODICITY, PERTURBATION, AND THE STRUCTURE OF ECOSYSTEMS

One basic question to be answered about any disturbance is How often does it occur? Fischer and Arthur (1977) found a periodicity of 32 million years in their oligotaxic and polytaxic episodes for the Meso-Cenozoic. They suggested that the mechanism was probably endogenic (perhaps related to sea-level change or magmatism) and not tied to solar or cosmic processes. The issue lay dormant until the publication of Alvarez et al.'s (1980) paper on impact and iridium at the K/T boundary. Later workers (such as Raup and Sepkoski, 1982; Sepkoski, 1989) then began to suggest, based on statistical analyses, that the periodicity was the result of periodic swarms of impacts, possibly caused by movement of our solar system through the central plane of the Milky Way Galaxy with a periodicity of 28–36 million years (Sepkoski, 1989, his figure 2). The gravitational effects would presumably dislodge asteroids and other bodies and increase the likelihood of their hitting the earth. Although the periodicity in the fossil record is said to be robust (Rampino and Caldeira, 1993), other workers have questioned its occurrence based on temporal control (Baksi, 1990) or new fossil data (Benton, 1995; see Rampino and Haggerty, 1995, for rebuttal). Part of the problem appears to be the massaging (filtering) of the data set before analysis (Rampino and Haggerty, 1995).

Almost without exception, investigations of mass extinction in the geologic record have focused on physical causes because physico-chemical phenomena are more easily resolved than short-term ecological processes in the rock record (chapters 4, 5). Perhaps this is also because from our own organismic (self-referential) perspective, it is "difficult for us to view processes at higher [ecosystem, biosphere] levels ... as emerging out of results of processes at lower biotic ... levels" (Salthe, 1985:219). Moreover, the evolution of higher levels (holons) may be incomplete. Upper-level systems may not be as highly evolved as lower-level ones because entification of higher levels proceeds at slower rates (Salthe, 1985; see also chapter 1).

The role of the structure and dynamics of ecological communities in determining the response of the community to perturbation is only now coming to be appreciated. For example, there may be some minimum recovery time required by a community before it reaches sufficient diversity to have another round of devastation recorded as an extinction in the fossil record (the rebound effect; McKinney, 1989; Stanley, 1990b). Using computer simulations, McKinney (1989) demonstrated that completely random external perturbations will produce a periodic signal in a logistic system (equation 5.3).

Moreover, the state of the communities themselves may determine the extent of their response (magnitude of extinction). In keeping with Allen and Starr's (1982) usage, the term *connectedness* is used here as a sort of summation of *connectivity* (the number of direct interconnections between one component and the rest of the system) and *connectance,* and refers to the number of interconnections via competition, predation, and other factors; connectedness is expressed as a percentage of the maximum number of interconnections. A common thread of much of the ecological literature until recently is that increased diversity results in increased stability (because of increased connectedness). As it turns out, this is not the case (May, 1973a, 1973b, 1976; Pimm, 1991). Systems that evolved during prolonged phases of environmental stability (such as prolonged greenhouses) were, in fact, quite diverse (Boucot, 1983, 1990), but they were probably also overconnected: There was both a loss of asymmetry and constraint in the hierarchy (high transitivity; Allen and Starr, 1982; Salthe, 1985; also see chapter 1). Holons with similar time constants are so strongly interconnected that any constraining signal (perturbation) is dclayed so long that the structure of the system has changed before the signal has had time to pass through the entire system. Signals either are lagged out of existence or become powerfully destabilizing because they elicit exactly the wrong response from the system by the time they have arrived to a part of the system. Thus the parts of the system are unable to act as a cohesive whole, and the system exhibits uncontrolled change (Allen and Starr, 1982). Perhaps this was the case for marine communities that had evolved over long periods of benign conditions before a major environmental perturbation (extinction) occurred, whether they were reefs adapted to oligotrophic conditions or "soft-bottom" communities that tended to occur in more nutrient-rich, productive waters (figure 5.4; Sheehan, 1985; see also Hallam, 1978).

Bear in mind, though, that minor disturbances can also elicit significant biotic response, especially in a highly overconnected system (McKinney, 1989; Plotnick and McKinney, 1993). If the data in Raup's (1991) histogram (his figure 4-4) are replotted as a natural log–log plot of the frequency of extinction versus the magnitude of extinction, as Stuart Kauffman (1993, 1995) has done (figure 8.2), we nearly get a straight-line relationship called a power "law" (but we need a few extinctions larger than those of the Permian to make the line straight).

In such a log–log relationship, the variables on each of the axes of the plot are related to each other by a straight line with a negative slope. In this particular case, there are many small extinctions and a few large ones. Similar relationships have been noted for earthquakes and floods of the Nile River (Kauffman, 1995; Bak, 1996). Note, however, that although a power law expresses a statistical relationship (like a gas law), it is ahistorical in nature: The actual causes of the relationship are immaterial and may vary considerably both qualitatively and quantitatively.

A power law can be produced in the laboratory by slowly building a pile of sand, one grain at a time. As each grain is added, there are frequent small avalanches (minor extinctions), but occasionally a large one occurs (mass extinction), even though only one grain is added at a time (Bak, 1996). This is what is called self-organized criticality. Many natural systems, including ecosystems, according to Kauffman (1993, 1995) and Bak (1996), evolve to a state poised between a highly inflexible region of order and one that is totally chaotic. In doing so, the system has enough flexibility to deal with perturbations while at the same time being able to adapt to change. However, "poised systems need no massive mover to move massively" (Kauffman, 1995:236). A minor agent of change is just as capable of causing a large avalanche as a small one. It all depends on the internal state of the system. This is precisely the behavior noted by Patterson and Fowler (1996) for extinction of planktic foraminiferal lineages.

Underconnected systems may also be highly unstable. There is no significant transfer of signal because the holons act at very different frequencies (figure 1.6). In this case there is no pathway that allows passage of sufficient signal between parts, and the parts act fairly independently of one another (Allen and Starr, 1982). Moreover, in simple model ecosystems disturbances may persist for thousands or tens of thousands of years; the time interval required for physical systems to approach asymptotic behavior (recovery) in similar experiments scales exponentially with the number of connections (Hastings and Higgins, 1994). Thus underconnected systems might characterize the time immediately following extinction, and may help to account for the delayed recovery of ecosystems (Hallam, 1991).

These concerns raise the larger issue of integration of a disturbance into a disturbed holon (Allen and Starr, 1982). The destructive power of a disturbance appears to be proportional to the difference in scale between the disturbance and the disturbed entity (Allen and Starr, 1982; Raup, 1994), but at the same time the scales (frequency) of the perturbation and the perturbed holon must be sufficiently similar for communication to occur (Allen and Starr, 1982; Salthe, 1985). Perturbations presumably cannot come from lower-level processes because these are the result of *averages* of large numbers of events (Salthe, 1985). For these reasons, perturbations must supposedly come from higher levels as distinct (perhaps even unique) events, or singularities; otherwise, they will be incorporated into the disturbed holon. (This idea seems to run counter to the preceding discussion of self-organized criticality and keystone species, so it is probably not just the number of connections but also the *strength* of the connections that matters.) Reice (1994) maintains that biological communities are almost always recovering from some sort of perturbation, and intermediate levels of disturbance (such as storms or disease) may actually promote biodiversity rather than lower it because disturbance prevents species from outcompeting and excluding one another (the intermediate distur-

FIGURE 8.4 Incorporation of disturbance into ecosystems. Note the inverse relationship between the frequency of disturbance and the magnitude of destruction. Beginning at time zero (compare with figure 1.6), repeated perturbations are progressively incorporated into ecosystems, and the perturbations become less destructive with time (cf. discussions of mass, minor, and background extinctions). *From T. F. H. Allen and T. B. Starr,* Hierarchy: Perspectives for Ecological Complexity. © *1982 by the University of Chicago.*

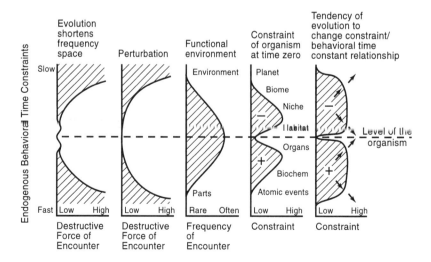

bance hypothesis; see chapter 5). In Reice's case we are dealing with small (and therefore frequent) perturbations to stream communities, which recover rapidly. These disturbances and the disturbed holon (stream community) act on similar (high-frequency) scales and the disturbance actually becomes part of the community (figure 8.4). Fire is another example. Although fire is considered a disturbance from the scale of an individual tree (it burns), fire has become incorporated into the dynamics of prairie and forest communities and is not a disturbance from the scale of tree species because it promotes germination of seedlings (Allen and Starr, 1982). In contrast, large disturbances that cause mass extinction occur on much larger spatial scales and much more infrequently, and so are unlikely to be incorporated into ecosystem dynamics (cf. figures 8.2 and 8.4).[1]

---

[1] P. H. Schultz and S. D'Hondt (1996, *Geology* 24: 963–967) have suggested that impact-related extinction is a function of both impact size and *angle*.

Epilogue: The Nature of Nature

*One of the principal objects of theoretical research . . . is to find the
point of view from which the subject appears in its greatest
simplicity. —J. Willard Gibbs*

Seek simplicity . . . and distrust it. —Alfred North Whitehead

*The present is the key to the past. —Geologic dictum based on the
writings of Charles Lyell*

The present may be the key to nothing at all. —Anonymous

I have attempted to accomplish a number of tasks in this book. For
one, I have attempted to show that what we measure in the stratigraphic record depends on the scale we use. There are undoubtedly
processes that are recognizable only in the context of geologic history
because they act at different levels with different rates that are otherwise imperceptible to us. It does make a difference, for example,
whether the dinosaurs died out with a bang or a whimper (Gould,
1992). Species are dying out with a bang right now because of the human race, but many of us, based on our species' limited perception
of time, do not perceive that anything is wrong. The geologic record
demonstrates that much larger—and virtually imperceptible—climate change can lead to sudden (even on geological timescales) climate change, not just at, say, the close of the Permian, but perhaps
eventually at the end of the Holocene because of us (Kerr, 1995a,
1995b). Of course, the geologic record of extinctions and other processes is imperfect, but perhaps taphonomic filters and stratigraphic
resolution would prove less problematic if we were more aware of the
limitations they impose on the rock record.

212

I have also argued, vehemently at times, for *why* history is important, not just from the viewpoint of an evolving planet but also from the viewpoint of how we study it. Casti (1994) emphasized that the scientific method is used in two distinct ways: to *predict* the outcome of future observation and to *explain* past observations. "Newton's laws of celestial motion are the quintessential example of . . . the first task, while the principle of natural selection in evolutionary biology exemplifies the second. Of course, in some cases the [scientific method] can be used for both purposes, as with Newton's laws of motion [predicting tides, the return of Halley's comet, and past and future eclipses, for example]. But this is generally not the case" (Casti, 1994:47). Predicting climate change is a prime example, one held up in the popular press almost every day, where historical science seems to fail. How can we accurately predict future climate when we do not completely understand the present (there may be millions of parameters and variables to be considered) and when natural systems are often nonlinear? And how can we predict the evolution of humankind based on evolutionary theory (Gould, 1986)? We cannot; therefore, in many scientists' eyes evolutionary biology and many other natural sciences are not genuinely "scientific." Moreover, the conflicting interpretations of past events and processes in the earth's history are viewed as evidence that geology is a "soft science," although conflicts abound in the "hard" sciences, as well (as in quantum mechanics).

Geology and biology *are* "scientific"; they are just not reductionist. We and our present surroundings are just the tip of an ancient iceberg. Much of natural systems (biological and physical) depends on what I will call conjunction (please ignore the astrological connotations), contingency (Gould, 1989), and constraint (which is really a part of contingency). The geologic record strongly indicates that much of what has occurred on this planet has often been the result of a unique (or nearly so) conjunction of circumstances (the Murder on the Orient Express hypothesis, for example). These circumstances in turn have determined what is to follow (in contingency, or a kind of "sensitive dependence on initial conditions"), except that the initial conditions are always changing through time and so the earth and its biota evolve, which gives us the irreversible "arrow of time" (Coveney and Highfield, 1990). Using uniformitarian arguments, causal ordering, and singularities (see Prologue), DeDuve (1995) deduced that even the metabolic pathways (chemical systems) of the lowliest cells are probably congruent with earlier prebiotic chemical pathways that eventually gave rise to life on earth. Contingency—*history*—constrains. No doubt it has been the same everywhere to the ends of the universe. Nevertheless, we continue to force ourselves to believe that in order to truly understand natural physical and biological systems we must describe them as simply as possible, meaning ahistorically (Frodeman, 1995), or that we must assume an almost absolute principle of uniformity (see Prologue).

Consider the Three-Ball Problem. In 1889, French mathematician Henri Poincaré demonstrated that although the behavior of two balls may be accurately predicted, the motion of three or more (such as Earth, the Sun, and the Moon) cannot (Coveney and Highfield, 1990; Ruelle, 1991). Their behavior is contingent on their previous behavior. Hence, the motions of the planets are only an approximation (perhaps the best we have in all of the sciences) because even though they are based on Newton's "laws," the planets appear to behave chaotically to a certain degree (Peterson, 1993).

As a result, in very complex physical systems (such as a volume of gas), which are still very simple compared to the cytoplasm of a lowly amoeba, laws (gas laws) used to describe or predict what will happen are *statistical* in nature (Coveney and Highfield, 1990; Ruelle, 1991). A canister of gas is an ergodic system; its components "explore" every possible configuration. Even if the exact position of molecules depends on their previous interactions with other molecules, the overall macroscopic behavior of the system can be described accurately, but only because the molecules explore every possible configuration quickly. Indeed, if Newton had truly been a reductionist, he would have reduced the behavior of the planets to simpler components (atoms). Instead, he simplified the system by using simpler (higher) entities called planets; so much for reductionist rhetoric (Cohen and Stewart, 1994).

The return time of a simple system, such as a canister of gas, to its current attractor (in this case, randomly dispersed molecules) is short. On the other hand, geological and biological systems have much longer return times (even though they may also be considered ergodic) because they explore the possible configurations more slowly. Thus equilibrium and steady-state are often figments of the laboratory. There are far too many parameters and variables acting at different levels and different frequencies (constraint) for the earth to behave otherwise; moreover, whether the holons are overconnected or underconnected may be crucial to their behavior. A signal transmitted through an underconnected system may generate a far different response from that in an overconnected one. Thus "history arises when the space of possibilities is too large by far for the actual to exhaust the possible" (Kauffman, 1995).

Considering the complexity of geological or biological systems, then, physics and chemistry look easy. No wonder so many biologists and geologists salivate at the thought of reductionist hyperbole applied to their own disciplines. No wonder some want to treat impacts as if they were molecular collisions ($PV = nRT$; cf. Figure 8.4) or that others wish to strip away the evolutionary permutations of life and reduce it to its presumed simplest (and in their eyes most important) component: DNA. Indeed, die-hard reductionists argue that living organisms are just special cases of physico-chemical systems (Casti, 1994). This view was common among some protozoologists of the late nineteenth and early twentieth centuries, who attempted to understand phenomena such as amoeboid motion by experimenting with fluids of different viscosities and densities (see Calkins, 1910, for brief review).

Historical scientists could just as easily say, but typically do not, that the shoe is on the other foot, and that physics and chemistry are really just special cases of biology (Casti, 1994). For example, the second law of thermodynamics states that the overall randomness or disorganization of our surroundings (entropy or heat death) has been steadily increasing since the universe was born. But this law is based on isolated (closed) laboratory settings, ones that do not exchange matter and energy with their surroundings (Josiah Willard Gibbs is one of the founders of modern equilibrium thermodynamics; Coveney and Highfield, 1990).

But galaxies, stars and planets, continents, and life have organized themselves and have evolved! Given that they are open thermodynamic systems, this is not surprising at all (Kauffman, 1993, 1995). And by evolving, these systems determined what was to follow. Many so-called laws are actually constrained by higher hierarchical levels, which are the boundary conditions or context of history (Allen and Starr, 1982; Eldredge and Salthe, 1984; Salthe, 1985; Cohen and Stewart, 1994). Perhaps what we call laws are, after all, merely "well-constructed Sherlock Holmes stories" (Cohen and Stewart, 1994:435): only the simplest forms of more complex rules that we still do not understand because they are not readily measured. This is not to say that some laws (or forces) do not act at high hierarchical levels (the law of gravity, for example), but others occur at much lower levels. Why is the molecular weight of chlorine, for example, 35.46 and not 37.50 or some other number? Because the molecular weight of chlorine is "a consequence of cosmology and history, not chemistry" (Cohen and Stewart, 1994:47). For that matter, if the cosmological constant, which is basically the amount of energy present in empty space (Freedman, 1996) were only slightly different, the universe as we know it would not even exist: If the constant were too large, it would act like "antigravity" and the universe would fly apart, and if too small the universe would collapse (Freedman, 1996). Some workers interpret this to mean that the fact that humankind even exists puts constraints on the properties of the universe that allow us to exist (the anthropic principle; Coveney and Highfield, 1990).

The principle of complementarity should reign. The so-called hard and soft (narrative) sciences are really two complementary ways of knowing (Frodeman, 1995). Reductionists seek to understand nature by looking at its insides, but we must also look at nature's outsides (Cohen and Stewart, 1994). Indeed, it was none other than Danish physicist Niels Bohr who, in 1929, coined the term *complementarity* to explain "physico-chemical processes on the one hand and the whole harmonious coordination . . . typical [of] life on the other" (Umbgrove, 1950:207).[1] Would we really understand organisms if we always reduced them to their component atoms and molecules (lower holons)? Do we really understand evolution by reducing it to an either/ or argument that pits gradualism against punctuationism or population versus organism versus gene? Would we really understand the solar system if we reduced it to the movement of the planets, and ignored its origin and evolution—and that of our galaxy—and their position (and ours) in the grand scheme of the universe? Would we really understand the earth and its life if we understood only the present? We and everything around us, as far as we can see, be it with an electron microscope or the Hubble telescope, are the products of history. The earth's history has been one long experiment, poorly constrained in a reductionist's eyes. How impoverished the earth would be if it had been otherwise.

---

[1] Umbgrove (1950) and Osborn (1918) are the only historical geology texts I have encountered that discuss this priciple or thermodynamics and life. Erwin Schrödinger's *What Is Life?* offers the viewpoint of a Nobel Laureate physicist. See also Brooks and Wiley (1988).

REFERENCES

Ahl, V., and T. F. H. Allen. 1996. *Hierarchy Theory: A Vision, Vocabulary, and Epistemology.* Columbia University Press, New York.

Algeo, T., R. A. Berner, J. B. Maynard, and S. E. Scheckler. 1993. Late Devonian oceanic anoxic events and biotic crises: "rooted" in the evolution of vascular plants? *GSA Today* 5(3).

Allen, T. F. H., and T. W. Hoekstra. 1992. *Toward a Unified Ecology.* Columbia University Press, New York.

Allen, T. F. H., and T. B. Starr. 1982. *Hierarchy: Perspectives for Ecological Complexity.* University of Chicago Press, Chicago.

Aller, R. C., and J. K. Cochran. 1976. ^{234}Th/^{238}Ur disequilibrium in near-shore sediment: particle reworking and diagenetic time scales. *Earth and Planetary Science Letters* 29:37–50.

Altabet, M. A., R. Francois, D. W. Murray, and W. L. Prell. 1995. Climate-related variations in denitrification in the Arabian Sea from sediment ^{15}N/^{14}N ratios. *Nature* 373:506–509.

Alvarez, L. W., W. Alvarez, F. Asaro, and H. V. Michel. 1980. Extraterrestrial cause for the Cretaceous–Tertiary extinction. *Science* 208:1095–1098.

Alvarez, W., P. Claeys, and S. W. Kieffer. 1995. Emplacement of Cretaceous–Tertiary boundary shocked quartz from Chicxulub crater. *Science* 269: 930–935.

Alve, E. 1995. Benthic foraminiferal response to estuarine pollution: a review. *Journal of Foraminiferal Research* 25:190–203.

Applin, E. R., A. E. Ellisor, and H. T. Kniker. 1925. Subsurface stratigraphy of the coastal plain of Texas and Louisiana. *American Association of Petroleum Geologists Bulletin* 9:79–122.

Armentrout, J. D. 1991. Paleontologic constraints on depositional modeling: examples of integration of biostratigraphy and seismic stratigraphy, Plio-Pleistocene, Gulf of Mexico. In P. Weimer and M. H. Link, eds., *Seismic Facies and Sedimentary Processes of Submarine Fans and Turbidite Systems*, pp. 137–170. Springer-Verlag, New York.

———. 1996. High-resolution sequence biostratigraphy: examples from the Gulf of Mexico Plio-Pleistocene. In J. A. Howell and J. F. Aitken, eds., *High Resolution Sequence Stratigraphy: Innovations and Applications.* Geological Society of London Special Publication No. 104:65–86.

Aronson, R. B. 1994. Scale-independent biological processes in the marine environment. *Oceanography and Marine Biology, Annual Review* 32:435–460.

Ausich, W. I., and D. J. Bottjer. 1991. History of tiering among suspension feeders in the benthic marine ecosystem. *Journal of Geological Education* 39:313–318.

Bak, P. 1996. *How Nature Works: The Science of Self-Organized Criticality.* Copernicus, New York.

Baksi, A. K. 1990. Search for periodicity in global events in the geologic record: quo vadimus? *Geology* 18:983–986.

Bambach, R. K. 1993. Seafood through time: changes in biomass, energetics, and productivity in the marine ecosystem. *Paleobiology* 19:372–397.

Bandy, O. L. 1953. Ecology and paleoecology of some California foraminifera. Part 1. The frequency distribution of Recent foraminifera off California. *Journal of Paleontology* 27:161–182.

Bandy, O. L., J. C. Ingle, and J. M. Resig. 1964a. Foraminiferal trends, Laguna Beach outfall area, California. *Limnology and Oceanography* 9:112–123.

———. 1964b. Foraminifera, Los Angeles County outfall area, California. *Limnology and Oceanography* 9:124–137.

———. 1965a. Foraminiferal trends, Hyperion outfall, California. *Limnology and Oceanography* 10:314–332.

———. 1965b. Modification of foraminiferal distribution by the Orange County outfall, California. *Ocean Science Engineering*, pp. 55–76.

Barber, R. T. 1992. Geological and climatic time scales of nutrient variability. In P. G. Falkowski and A. D. Woodhead, eds., *Primary Productivity and Biogeochemical Cycles in the Sea*, pp. 89–106. Plenum, New York.

Barmawidjaja, D. M., G. J. van der Zwann, F. J. Jorissen, and S. Puskaric. 1995. 150 years of eutrophication in the northern Adriatic Sea: evidence from a benthic foraminiferal record. *Marine Geology* 122:367–384.

Barrell, N.J. 1917. Rhythms and the measurements of geologic time. *Geological Society of America Bulletin* 28:745–904.

Barrera, E. 1994. Global environmental changes preceding the Cretaceous–Tertiary boundary: early–late Maastrichtian transition. *Geology* 22:877–880.

Barron, J. A. 1993. Diatoms. In J. H. Lipps, ed., *Fossil Prokaryotes and Protists*, pp. 155–167. Blackwell Scientific, Boston.

Bates, N. R., and U. Brand. 1990. Secular variation of calcium carbonate mineralogy; an evaluation of ooid and micrite chemistries. *Geologische Rundschau* 79:27–46.

Bazzaz, F., and E. D. Fajer. 1992. Plant life in a CO_2-rich world. *Scientific American* (January):68–74.

Beck, R. A., D. W. Burbank, W. J. Sercombe, T. L. Olson, and A. M. Khan. 1995. Organic carbon exhumation and global warming during the early Himalayan collision. *Geology* 23:387–390.

Behrensmeyer, A. K., and S. M. Kidwell. 1985. Taphonomy's contributions to paleobiology. *Paleobiology* 11:105–119.

Beier, J. A., and J. M. Hayes. 1989. Geochemical and isotopic evidence for paleoredox conditions during deposition of the Devonian-Mississippian New Albany Shale, southern Indiana. *Geological Society of America Bulletin* 101:774–782.

Bell, P. R. F. 1991. Status of eutrophication in the Great Barrier Reef Lagoon. *Marine Pollution Bulletin* 23:89–93.

——. 1992. Eutrophication and coral reef: some examples in the Great Barrier Reef Lagoon. *Water Research* 26:553–568.

Beltrami, E. 1993. *Mathematical Models in the Social and Biological Sciences.* Jones and Bartlett, Boston.

Benninger, L. K., R. C. Aller, J. K. Cochran, and K. K. Turekian. 1979. Effects of biological sediment mixing on the ^{210}Pb chronology and trace metal distribution in a Long Island Sound sediment core. *Earth and Planetary Science Letters* 43:241–259.

Benton, M. J. 1979. Increase in total global biomass over time. *Evolutionary Theory* 4:123–128.

——. 1995. Diversification and extinction in the history of life. *Science* 268:52–58.

Berg, H. C. 1993. *Random Walks in Biology.* Princeton University Press, Princeton, New Jersey.

Berger, W. H. 1977. Carbon dioxide excursions and the deep-sea record: aspects of the problem. In N. R. Andersen and A. Malahoff, eds., *The Fate of Fossil Fuel CO_2 in the Oceans*, pp. 505–542. Plenum, New York.

Berger, W. H., A. A. Ekdale, and P. P. Bryant. 1979. Selective preservation of burrows in deep-sea carbonates. *Marine Geology* 32:205–230.

Berger, W. H., and G. R. Heath. 1968. Vertical mixing in pelagic sediments. *Journal of Marine Research* 26:134–143.

Berger, W. H., and J. S. Killingley. 1982. Box cores from the equatorial Pacific: [14]C sedimentation rates and benthic mixing. *Marine Geology* 45:93–125.

Berger, W. H., and A. Soutar. 1970. Preservation of plankton shells in an anaerobic basin off California. *Geological Society of America Bulletin* 81: 275–282.

Berner, R. A. 1971. *Principles of Chemical Sedimentology.* McGraw-Hill, New York.

——. 1989. Biogeochemical cycles of carbon and sulfur and their effect on atmospheric oxygen over Phanerozoic time. *Palaeogeography, Palaeoclimatology, Palaeoecology (Global and Planetary Change Section)* 75:97–122.

———. 1990. Atmospheric carbon dioxide levels over Phanerozoic time. *Science* 249:1382–1386.

———. 1992. Weathering, plants, and the long-term carbon cycle. *Geochimica et Cosmochimica Acta* 56:3225–3231.

———. 1993. Paleozoic atmospheric CO_2: importance of solar radiation and plant evolution. *Science* 261:68–70.

Berner, R. A., and A. Lasaga. 1989. Modeling the geochemical carbon cycle. *Scientific American* 60(3):74–81.

Berner, R. A., A. Lasaga, and R. M. Garrels. 1983. The carbonate–silicate geochemical cycle and its effect on atmospheric carbon dioxide over the past 100 million years. *American Journal of Science* 283:641–683.

Berner, R. A., and R. Raiswell. 1983. Burial of organic carbon and pyrite sulfur in sediments over Phanerozoic time: a new theory. *Geochimica et Cosmochimica Acta* 4:855–862.

Berry, W. B. N. 1987. *Growth of a Prehistoric Time Scale: Based on Organic Evolution.* Blackwell, Palo Alto, Calif.

Berry, W. B. N., and P. Wilde. 1978. Progressive ventilation of the oceans: an explanation for the distribution of the lower Paleozoic black shales. *American Journal of Science* 278:257–275.

Berry, W. B. N., P. Wilde, and M. S. Quinby-Hunt. 1987. The oceanic non-sulfidic oxygen minimum zone: a habitat for graptolites? *Bulletin of the Geological Society of Denmark* 35:103–114.

Boggs, S. 1987. *Principles of Sedimentology and Stratigraphy.* Merrill, Columbus, Ohio.

Bond, G. C. 1995. Climate and the conveyor. *Nature* 377:383–384.

Bond, G. C., and R. Lotti. 1995. Iceberg discharges into the North Atlantic on millennial time scales during the last glaciation. *Science* 267:1005–1010.

Bondi, H. 1994. Karl Popper (1902–1994). *Nature* 371:478.

Boss, S. K., and B. H. Wilkinson. 1991. Planktogenic/eustatic control on cratonic oceanic carbonate accumulation. *Journal of Geology* 99:497–513.

Bottjer, D. J., K. A. Campbell, J. A. Schubert, and M. L. Droser. 1995. Palaeoecological models, non-uniformitarianism, and tracking the changing ecology of the past. In D. W. J. Bosence and P. A. Allison, eds., *Marine Palaeoenvironmental Analysis from Fossils,* Geological Society of London Special Publication 83:7–26.

Bottjer, D. J., and M. L. Droser. 1994. The history of Phanerozoic bioturbation. In S. K. Donovan, ed., *The Palaeobiology of Trace Fossils,* pp. 155–176. Wiley, Chichester.

Boucot, A. J. 1983. Does evolution take place in an ecological vacuum? *Journal of Paleontology* 57:1–30.

———. 1990. Community evolution: Its evolutionary and biostratigraphic significance. In W. Miller, ed., *Paleocommunity Temporal Dynamics: The Long-*

Term Development of Multispecies Assemblies. Paleontological Society Special Publication 5:48–70.

Boudreau, B. P. 1986a. Mathematics of tracer mixing in sediments: I. Spatially dependent, diffusive mixing. *American Journal of Science* 286:161–198.

———. 1986b. Mathematics of tracer mixing in sediments: II. Nonlocal mixing and biological conveyor-belt phenomena. *American Journal of Science* 286: 199–238.

Boudreau, B. P., and D. M. Imboden. 1987. Mathematics of tracer mixing in sediments: III. The theory of nonlocal mixing within sediments. *American Journal of Science* 287:693–719.

Boutilier, R. G., T. G. West, G. H. Pogson, K. A. Mesa, J. Wells, and M. J. Wells. 1996. Nautilus and the art of metabolic maintenance. *Nature* 382:534–536.

Boyle, E. A., and L. D. Keigwin. 1982. Deep circulation of the North Atlantic over the last 200,000 years: geochemical evidence. *Science* 218:784–787.

———. 1987. North Atlantic thermohaline circulation during the past 20,000 years linked to high-latitude surface temperature. *Nature* 330:35–40.

Bralower, H. J., and J. R. Thierstein. 1984. Low productivity and slow deep-water circulation in mid-Cretaceous oceans. *Geology* 12:614–618.

Bramlette, M. N. 1958. Significance of coccoliths in calcium carbonate deposition. *Geological Society of America Bulletin* 69:121–126.

———. 1965. Massive extinctions in biota at the end of Mesozoic time. *Science* 148:1696–1699.

Brasier, M. 1986. Why do lower plants and animals biomineralize? *Paleobiology* 12:241–250.

Brasier, M. D. 1992. Paleoceanography and changes in the biological cycling of phosphorus across the Precambrian-Cambrian boundary. In J. H. Lipps and P. W. Bengston, eds., *Origin and Early Evolution of the Metazoa,* pp. 483–523. Plenum, New York.

———. 1995. Fossil indicators of nutrient levels. 2: evolution and extinction in relation to oligotrophy. In D. W. J. Bosence and P. A. Allison, eds., *Marine Paleoenvironmental Analysis from Fossils.* Geological Society of London Special Publication No. 83:133–150.

Brasier, M. D., R. M. Corfield, L. A. Derry, A. Y. Rozanov, and A. Y. Zhuravlev. 1994. Multiple $\delta^{13}C$ excursions spanning the Cambrian explosion to the Botomian crisis in Siberia. *Geology* 22:455–458.

Brass, G. W., J. R. Southam, and W. H. Peterson. 1982. Warm saline bottom water in the ancient ocean. *Nature* 296:620–623.

Bremer, M. L., and G. P. Lohmann. 1982. Evidence for primary control of the distribution of certain Atlantic Ocean benthonic foraminifera by degree of carbonate saturation. *Deep-Sea Research* 29:987–998.

Brenchley, P. J. 1989. The late Ordovician extinction. In S. K. Donovan, ed., *Mass Extinctions: Processes and Evidence,* pp. 104–132. Columbia University Press, New York.

Brenchley, P. J., J. D. Marshall, G. A. F. Carden, D. B. R. Roberston, D. G. F.

Long, T. Meidla, L. Hints, and T. F. Anderson. 1994. Bathymetric and isotopic evidence for a short-lived Late Ordovician glaciation in a greenhouse period. *Geology* 22:295–298.

Bresler, V., and V. Yanko. 1995. Chemical ecology: a new approach to the study of living benthic epiphytic foraminifera. *Journal of Foraminiferal Research* 25:267–279.

Bretsky, P. W., and S. M. Klofak. 1986. "Rules of assembly" for two Late Ordovician communities. *Palaios* 1:462–477.

Brett, C. E., and G. C. Baird. 1986. Comparative taphonomy: a key to paleoenvironmental interpretation based on fossil preservation. *Palaios* 3:207–227.

Briggs, J. C. 1994. Species diversity: land and sea compared. *Systematic Biology* 43:130–135.

Broecker, W. S. 1982. Ocean chemistry during glacial time. *Geochimica et Cosmochimica Acta* 46:1689–1705.

——. 1987. Unpleasant surprises in the greenhouse? *Nature* 328:123–126.

——. 1995. Chaotic climate. *Scientific American* 273(11):62–68.

——. 1997. Will our ride into the greenhouse future be a smooth one? *GSA Today* 7(5):1–7.

Broecker, W. S., and T.-H. Peng. 1982. *Tracers in the Sea.* Eldigio Press (Lamont-Doherty Geological Observatory), Palisades, New York.

——. 1989. The cause of the glacial to interglacial atmospheric CO_2 change: a polar alkalinity hypothesis. *Global Biogeochemical Cycles* 3:215–239.

Broecker, W. S., D. M. Peteet, and D. Rind. 1985. Does the ocean-atmosphere system have more than one stable mode of operation? *Nature* 315:21–26.

Brooks, D. R., and E. O. Wiley. 1988. *Evolution as Entropy.* University of Chicago Press, Chicago.

Burlando, B. 1993. The fractal geometry of evolution. *Journal of Theoretical Biology* 163:161–172.

Busch, R. M., and R. R. West. 1987. Hierarchical genetic stratigraphy: a framework for paleoceanography. *Paleoceanography* 2:141–164.

Buzas, M. A., and S. J. Culver. 1994. Species pool and dynamics of marine paleocommunities. *Science* 264:1439–1441.

Calkins, G. N. 1910. *The Protozoa.* Macmillan, London.

Canfield, D. E. 1991. Sulfate reduction in deep-sea sediments. *American Journal of Science* 291:177–188.

Canfield, D. E., and R. Raiswell. 1991. Carbonate precipitation and dissolution: Its relevance to fossil preservation. In P. A. Allison and D. E. G. Briggs, eds, *Taphonomy: Releasing the Data Locked in the Fossil Record,* pp. 411–453. Plenum, New York.

Canfield, D. E., and A. Teske. 1996. Late Proterozoic rise in atmospheric oxygen concentration inferred from phylogenetic and sulfur-isotope studies. *Nature* 382:127–132.

Caplan, M. L., R. M. Bustin, and K. A. Grimm. 1996. Demise of a Devonian–Carboniferous carbonate ramp by eutrophication. *Geology* 24:715–718.

Carslaw, H. S., and J. C. Jaeger. 1959. *Conduction of Heat in Solids.* Oxford

University Press, London.

Casey, R. E. 1993. Radiolaria. In J. H. Lipps, ed., *Fossil Prokaryotes and Protists*, pp. 249–284. Blackwell Scientific, Boston.

Casti, J. L. 1994. *Complexification: Explaining a Paradoxical World Through the Science of Surprise.* HarperCollins, New York.

Causton, D. R. 1987. *A Biologist's Advanced Mathematics.* Allen and Unwin, London.

Claeys, P., J.-G. Casier, and S. V. Margolis. 1992. Microtektites and mass extinctions: evidence for a Late Devonian asteroid impact. *Science* 257:1102–1104.

Clark, D. L. 1987. Phylum Conodonta. In R. S. Boardman, A. H. Cheetham, and A. J. Rowell, eds., *Fossil Invertebrates,* pp. 636–662. Blackwell, Palo Alto, Calif.

Claypool, G. E., W. T. Holser, I. R. Kaplan, H. Sakai, and I. Zak. 1980. The age curves of sulfur and oxygen isotopes in marine sulfate and their mutual interpretation. *Chemical Geology* 28:199–260.

Cloetingh, S. 1988. Intraplate stresses: a tectonic cause for third-order cycles in apparent sea level? In C. K. Wilgus, B. S. Hastings, C. G. Kendall, H. W. Posamentier, C. A. Ross, and J. C. Van Wagoner, eds., *Sea-Level Changes: An Integrated Approach.* Tulsa, Oklahoma. Society of Economic Paleontologists and Mineralogists Special Publication No. 42:19–29.

CoBabe, E. A., and W. D. Allmon. 1994. Effects of sampling on paleoecologic and taphonomic analysis in high diversity fossil accumulations: an example from the Eocene Gosport Sand, Alabama. *Lethaia* 27:167–178.

Cohen, J., and I. Stewart. 1994. *The Collapse of Chaos: Discovering Simplicity in a Complex World.* Viking, New York.

Compton, J. S., D. A. Hodell, J. R. Garrido, and D. J. Mallinson. 1993. Origin and age of phosphorite from the south-central Florida Platform. Relation of phosphogenesis to sea-level fluctuations and $\delta^{13}C$ excursions. *Geochimica et Cosmochimica Acta* 57:131–146.

Connell, J. H. 1978. Diversity in tropical rainforests and coral reefs. *Science* 199:1302–1310.

Connell, J. H., and R. O. Slatyer. 1977. Mechanisms of succession in natural communities and their role in community stability and organization. *American Naturalist* 111:1119–1144.

Cook, P. J., and J. H. Shergold. 1984. Phosphorus, phosphorites and skeletal evolution at the Precambrian-Cambrian boundary. *Nature* 308:231–236.

Cooper, S. L. 1995. Chesapeake Bay watershed historical land use: impact on water quality and diatom communities. *Ecological Applications* 5:703–723.

Copper, P. 1974. Structure and development of early Paleozoic reefs. Proceedings, 2nd International Coral Reef Symposium 1:365–386.

——. 1988. Ecological succession in Phanerozoic reef ecosystems: is it real? *Palaios* 3:136–152.

Corliss, B. H. 1985. Microhabitats of benthic foraminifera within deep-sea sediments. *Nature* 314:435–438.

REFERENCES

——. 1991. Morphology and microhabitat preferences of benthic foraminifera from the northwest Atlantic Ocean. *Marine Micropaleontology* 17:195–236.

Corliss, B. H., and S. Emerson. 1990. Distribution of rose bengal stained deep-sea benthic foraminifera from the Nova Scotian continental margin and Gulf of Maine. *Deep-Sea Research* 37:381–400.

Corliss, B. H., and S. Honjo. 1981. Dissolution of deep-sea benthonic foraminifera. *Micropaleontology* 27:356–378.

Courtillot, V., and Y. Gaudemer. 1996. Effects of mass extinctions on biodiversity. *Nature* 381:146–148.

Coveney, P., and R. Highfield. 1990. *The Arrow of Time.* Fawcett Columbine, New York.

Cracraft, J. 1985. Species selection, macroevolutionary analysis, and the "hierarchical theory" of evolution. *Systematic Zoology* 34:222–229.

Crank, J. 1975. *The Mathematics of Diffusion.* Oxford University Press, London.

Crowley, T. J., and S. K. Baum. 1991. Towards reconciliation of Late Ordovician (~440 Ma) glaciation with very high CO_2 levels. *Journal of Geophysical Research* 96:22,597–22,610.

Crowley, T. J., and G. R. North. 1988. Abrupt climate change and extinction events in Earth history. *Science* 240:996–1002.

Culver, S. J. 1990. Benthic foraminifera of Puerto Rican mangrove-lagoon systems: Potential for paleoenvironmental interpretations. *Palaios* 5:34–51.

——. 1991. Early Cambrian foraminifera from west Africa. *Science* 254:689–691.

Culver, S. J., and M. A. Buzas. 1994. Species pool and dynamics of marine paleocommunities. *Science* 264:1439–1441.

——. 1995. The effects of anthropogenic habitat disturbance, habitat destruction, and global warming on shallow marine benthic foraminifera. *Journal of Foraminiferal Research* 25:204–211.

Cushman, J. A. 1921. Foraminifera from the north coast of Jamaica. *United States National Museum Proceedings* 59:47–82.

Cutler, A. H. 1993. Mathematical models of temporal mixing in the fossil record. In S. M. Kidwell and A. K. Behrensmeyer, eds., *Taphonomic Approaches to Time Resolution in Fossil Assemblages.* Paleontological Society Short Courses in Paleontology No. 6:169–187.

Cutler, A. H., and K. W. Flessa. 1990. Fossils out of sequence: computer simulations and strategies for dealing with stratigraphic disorder. *Palaios* 5:227–235.

Dao-Yi, X., and Y. Zheng. 1993. Carbon isotope and iridium event markers near the Permian/Triassic boundary in the Meishan section, Zhejiang Province, China. *Palaeogeography, Palaeoclimatology, Palaeoecology* 104:171–176.

Darwin, C. 1859. *On the Origin of Species.* Murray, London.

Davidson, E. A., D.C. Nepstad, C. Klink, and S. E. Trumbore. 1995a. Pasture soils as carbon sinks. *Nature* 376:472–473.

Davidson, E. H., K. J. Peterson, and R. A. Cameron. 1995b. Origin of bilaterian body plans: evolution of developmental regulatory mechanisms. *Science* 270:1319–1325.

Davis, J. C. 1986. *Statistics and Data Analysis in Geology.* Wiley, New York.

DeAngelis, D. L. 1992. Dynamics of nutrient cycling and food webs. Chapman and Hall, London.

DeDuve, C. 1995. *Vital Dust: Life as a Cosmic Imperative.* Basic Books, New York.

Denne, R. A. 1994. Operational applications of graphic correlation. In H. R. Lane, G. Blake, and N. R. MacLeod, eds., *Graphic Correlation and the Composite Standard: The Methods and their Applications.* Society for Sedimentary Geology Research Conference, November 27–December 2, Houston, Tex.

Denne, R. A., and B. K. Sen Gupta. 1989. Effects of taphonomy and habitat on the record of benthic foraminifera in modern sediments. *Palaios* 4:414–423.

deRuiter, P. C., A.-M. Neutel, and J. C. Moore. 1995. Energetics, patterns of interaction strengths, and stability in real ecosystems. *Science* 269:1257–1260.

DiMichele, W. A. 1994. Ecological patterns in time and space. *Paleobiology* 20:89–92.

Dingus, L. 1984. Effects of stratigraphic completeness on interpretations of extinction rates across the Cretaceous–Tertiary boundary. *Paleobiology* 10:420–438.

Dingus, L., and P. M. Sadler. 1982. The effects of stratigraphic completeness on estimates of evolutionary rates. *Systematic Zoology* 31:400–412.

Dodd, J. R., and R. J. Stanton. 1981. *Paleoecology, Concepts and Applications.* Wiley, New York.

Done, T. J. 1992. Phase shifts in coral reef communities and their ecological significance. *Hydrobiologia* 247:121–132.

Donovan, S. K. 1989a. Palaeontological criteria for the recognition of mass extinction. In S. K. Donovan, ed., *Mass Extinctions: Processes and Evidence,* pp. 19–36. Columbia University Press, New York.

——, ed. 1989b. *Mass Extinctions: Processes and Evidence.* Columbia University Press, New York.

Doolittle, R. F., D.-F. Feng, S. Tsang, G. Cho, and E. Little. 1996. Determining divergence times of the major kingdoms of living organisms with a protein clock. *Science* 271:470–477.

Douglas, R. G., J. Liestman, C. Walch, G. Blake, and M. L. Cotton. 1980. The transition from live to sediment assemblage in benthic foraminifera from the southern California borderland. In M. E. Field, A. H. Bouma, I. P. Colburn, R. G. Douglas, and J. C. Ingle, eds., *Quaternary Depositional Environments of the Pacific Coast.* Society of Economic Paleontologists and Mineralogists, Pacific Section 4:257–280.

Dunbar, C. O., and J. Rogers. 1957. *Principles of Stratigraphy.* Wiley, New York.

Dyer, B. D., and R. A. Obar. 1994. *Tracing the History of Eukaryotic Cells: The Enigmatic Smile.* Columbia University Press, New York.

REFERENCES

Edinger, E. N., and M. J. Risk. 1994. Oligo-Miocene extinction and geographic restriction of Caribbean corals: roles of turbidity, temperature, and nutrients. *Palaios* 9:576–598.

——. 1995. Preferential survivorship of brooding corals in a regional extinction. *Paleobiology* 21:200–219.

Edwards, L. E. 1989a. Quantitative biostratigraphy. In N. L. Gilinsky and P. W. Signor, eds., *Analytical Paleobiology.* Paleontological Society Short Courses in Paleontology No. 4:39–58.

——. 1989b. Supplemented graphic correlation: a powerful tool for paleontologists and non-paleontologists. *Palaios* 4:127–143.

Eicher, D. L. 1976. *Geologic Time.* Prentice-Hall, Englewood Cliffs, N.J.

Eisner, T., J. Lubchenco, E. O. Wilson, D. S. Wilcove, and M. J. Bean. 1995. Building a scientifically sound policy for protecting endangered species. *Science* 269:1231–1232.

Ekdale, A. A., R. G. Bromley, and S. G. Pemberton. 1984. Ichnology: the use of trace fossils in sedimentology and stratigraphy. Society of Economic Paleontologists and Mineralogists Short Course Number 15, Tulsa, Okla.

Eldredge, N. 1991. *The Miner's Canary: Unraveling the Mysteries of Extinction.* Prentice-Hall, Englewood Cliffs, N.J.

——. 1992. *Systematics, Ecology, and Biodiversity.* Columbia University Press, New York.

——. 1995. *Reinventing Darwin: The Great Debate at the High Table of Evolutionary Theory.* Wiley, New York.

Eldredge, N., and S. J. Gould. 1972. Punctuated equilibria: an alternative to phyletic gradualism. In T. J. M. Schopf, ed., *Models in Paleobiology,* pp. 82–115. W. H. Freeman, San Francisco.

Eldredge, N., and S. N. Salthe. 1984. Hierarchy and evolution. *Oxford Surveys in Evolutionary Biology* 1:182–206.

Enos, P. 1991. Sedimentary parameters for computer modeling. In E. K. Franseen, W. L. Watney, C. G. Kendall, and W. Ross, eds., *Sedimentary Modeling: Computer Simulations and Methods for Improved Parameter Definition. Kansas Geological Survey Bulletin* 233:63–99.

Erwin, D. H. 1993. *The Great Paleozoic Crisis: Life and Death in the Permian.* Columbia University Press, New York.

Erwin, D. H., J. Valentine, and D. Jablonski. 1997. The origin of animal body plans. *American Scientist* 85:126–137.

Filippelli, G. M., and M. L. Delaney. 1992. Similar phosphorus fluxes in ancient phosphorite deposits and a modern phosphogenic environment. *Geology* 20:709–712.

——. 1994. The oceanic phosphorus cycle and continental weathering during the Neogene. *Paleoceanography* 9:643–652.

Fischer, A. G. 1984. The two Phanerozoic supercycles. In W. A. Berggren and J. A. van Couvering, eds., *Catastrophes and Earth History,* pp. 129–150. Princeton University Press, Princeton, N.J.

Fischer, A. G., and M. A. Arthur. 1977. Secular variations in the pelagic realm.

In H. E. Cook and P. Enos, eds., *Deep-Water Carbonate Environments.* Society of Economic Paleontologists and Mineralogists Special Publication 25:19–50.

Fisher, M. J., I. M. Rao, M. A. Ayarza, C. E. Lascano, J. I. Sanz, R. J. Thomas, and R. R. Vera. 1994. Carbon storage by introduced deep-rooted grasses in the South American savannas. *Nature* 371:236–238.

Flessa, K. W. 1990. The "facts" of mass extinctions. In V. L. Sharpton and P. D. Ward, eds., *Global Catastrophes in Earth History: An Interdisciplinary Conference on Impacts, Volcanism, and Mass Mortality.* Geological Society of America Special Paper 247.

Flessa, K. W., Cutler, A. H., and K. H. Meldahl. 1993. Time and taphonomy: quantitative estimates of time-averaging and stratigraphic disorder in a shallow marine habitat. *Paleobiology* 19:266–286.

Flessa, K. W., and D. Jablonski. 1985. Declining Phanerozoic background extinction rates: effect of taxonomic structure? *Nature* 313:216–218.

Flessa, K. W., and M. Kowalewski. 1994. Shell survival and time-averaging in nearshore and shelf environments: estimates from the radiocarbon literature. *Lethaia* 27:153–165.

Föllmi, K. B., H. Weissert, and A. Lini. 1993. Nonlinearities in phosphogenesis and phosphorus-carbon coupling and their implications for global change. In R. Wollast, F. T. Mackenzie, and L. Chou, eds., *Interactions of C, N, P and S Biogeochemical Cycles and Global Change,* pp. 447–474. Springer-Verlag, Berlin.

Fox, R. F. 1988. *Energy and the Evolution of Life.* W. H. Freeman, New York.

Frakes, L. A., J. E. Francis, and J. I. Syktus. 1992. *Climate Modes of the Phanerozoic.* Cambridge University Press, Cambridge, England.

François, L. M., and J. C. G. Walker. 1992. Modelling the Phanerozoic carbon cycle and climate: constraints from the $^{87}Sr/^{87}Sr$ isotopic ratio of seawater. *American Journal of Science* 292:81–135.

Freedman, D. H. 1996. The mediocre universe. *Discover* 17(2):64–75.

Frodeman, R. 1995. Geological reasoning: geology as an interpretive and historical science. *Geological Society of America Bulletin* 107:960–968.

Fürsich, F. T. 1978. The influence of faunal condensation and mixing on the preservation of fossil benthic communities. *Lethaia* 11:243–250.

Futuyama, D. J., and Moreno. 1988. The evolution of ecological specialization. *Annual Review of Ecology and Systematics* 19:207–233.

Ganeshram, R. S., T. F. Pedersen, S. E. Calvert, and J. W. Murray. 1995. Large changes in oceanic nutrient inventories from glacial to interglacial periods. *Nature* 376:755–758.

Garrels, R. M., A. Lerman, and F. T. Mackenzie. 1976. Controls of atmospheric O_2 and CO_2: past, present, and future. *American Scientist* 64:306–315.

Gibbons, A. 1996. On the many origins of species. *Science* 273:1496–1499.

Glass, B. P. 1969. Reworking of deep-sea sediments as indicated by the vertical dispersion of the Australasian and Ivory Coast microtektite horizons. *Earth and Planetary Science Letters* 6:409–415.

REFERENCES

Glass, B. P., and J. E. Hazel. 1990. Chronostratigraphy of Upper Eocene microspherules: comment & reply. *Palaios* 5:387–390.

Goldstein, S. T., and R. W. Frey. 1986. Salt marsh foraminifera, Sapelo Island, Georgia. *Senckenbergiana Maritima* 18:97–121.

Goldstein, S. T., and E. B. Harben. 1993. Taphofacies implications of infaunal foraminiferal assemblages in a Georgia salt marsh, Sapelo Island. *Micropaleontology* 39:53–62.

Goldstein, S. T., G. T. Watkins, and R. M. Kuhn. 1995. Microhabitats of salt marsh foraminifera, St. Catherine's Island, Georgia, U.S.A. *Marine Micropaleontology* 26:17–29.

Goodwin, P. W., and E. J. Anderson. 1985. Punctuated aggradational cycles: A general hypothesis of episodic stratigraphic accumulation. *Journal of Geology* 93:515–533.

Gould, S. J. 1986. Evolution and the triumph of homology, or why history matters. *American Scientist* 74:60–69.

——. 1989. *Wonderful Life: The Burgess Shale and the Nature of History.* W. W. Norton, New York.

——. 1992. Dinosaurs in the haystack. *Natural History* (March):2–12.

——. 1996. Creating the creators. *Discover* (October) 17:42–54.

Graham, J. B., R. Dudley, N. M. Aguilar, and C. Gans. 1995. Implications of the late Palaeozoic oxygen pulse for physiology and evolution. *Nature* 375:117–120.

Green, M. A., R. C. Aller, and J. Y. Aller. 1993a. Experimental evaluation of the influences of biogenic reworking on carbonate preservation in nearshore sediments. *Marine Geology* 107:175–181.

——. 1993b. Carbonate dissolution and temporal abundances of foraminifera in Long Island Sound sediments. *Limnology and Oceanography* 38:331–345.

Greene, M. T. 1982. *Geology in the Nineteenth Century: Changing Views of a Changing World.* Cornell University Press, Ithaca, N.Y.

Greenstein, B. J., J. M. Pandolfi, and P. J. Moran. 1995. Taphonomy of crown-of-thorns starfish: implications for recognizing ancient population outbreaks. *Coral Reefs* 14:91–97.

Guinasso, N. L., and D. R. Schink. 1975. Quantitative estimates of biological mixing rates in abyssal sediments. *Journal of Geophysical Research* 80:3032–3043.

Hallam, A. 1978. How rare is phyletic gradualism? Evidence from Jurassic bivalves. *Paleobiology* 4:16–29.

——. 1981. *Facies Interpretation and the Stratigraphic Record.* W. H. Freeman, San Francisco, Calif.

——. 1989. The case for sea-level change as a dominant causal factor in mass extinction of marine invertebrates. *Philosophical Transactions of the Royal Society of London* B325:437–455.

——. 1991. Why was there a delayed radiation after the end-Palaeozoic extinctions? *Historical Biology* 5:257–262.

———. 1992. *Phanerozoic Sea-Level Changes.* Columbia University Press, New York.

Hallock, P. 1982. Evolution and extinction in larger foraminifera. *Proceedings of the Third North American Paleontological Convention* 1:221–225.

———. 1987. Fluctuations in the trophic resource continuum: a factor in global diversity cycles? *Paleoceanography* 2:457–471.

Hallock, P., F. E. Müller-Karger, and J. C. Halas. 1993. Coral reef decline. *National Geographic Exploration and Research* 9:358–378.

Hallock, P., I. Premoli Silva, and A. Boersma. 1991. Similarities between planktonic and larger foraminiferal evolutionary trends through Paleogene paleoceanographic changes. *Palaeogeography, Palaeoclimatology, Palaeoecology* 83:49–64.

Hallock, P., and W. Schlager. 1986. Nutrient excess and the demise of coral reefs and carbonate platforms. *Palaios* 1:389–398.

Hallock, P., H. Talge, and B. Cockey. 1995. Larger foraminifera as environmental indicators for coral reefs. *Congress Program and Abstracts.* SEPM Congress on Sedimentary Geology, St. Petersburg, Fla.

Haq, B. U., J. Hardenbol, and P. R. Vail. 1988. Mesozoic and Cenozoic chronostratigraphy and cycles of sea-level change. In C. K. Wilgus, B. S. Hastings, C. G. Kendall, H. W. Posamentier, C. A. Ross, and J. C. Van Wagoner, eds., *Sea-Level Changes: An Integrated Approach.* Tulsa, Oklahoma. Society of Economic Paleontologists and Mineralogists Special Publication No. 42:71–108.

Harper, H. E., and A. H. Knoll. 1975. Silica, diatoms, and Cenozoic radiolarian evolution. *Geology* 3:175–177.

Harrison, S. L. 1996. Do taxa persist as metapopulations in evolutionary time? In M. L. McKinney and J. A. Drake, eds., *Biodiversity Dynamics: Origination and Extinction of Populations, Species, Communities, and Higher Taxa.* Columbia University Press, New York. In press.

Hart, M. B. 1996. The geology and micropaleontology of the Channel tunnel. In J. E. Repetski, ed., *Sixth North American Paleontological Convention, Abstracts and Program,* p. 162. Paleontological Society Special Publication No. 8.

Hastings, A., and K. Higgins. 1994. Persistence of transients in spatially structured ecological models. *Science* 263:1133–1136.

Herbert, T. D., and J. L. Sarmiento. 1991. Ocean nutrient distribution and oxygenation: Limits on the formation of warm saline bottom water over the past 91 m. y. *Geology* 19:702–705.

Hippensteel, S. P., and R. E. Martin. 1998. Foraminifera as indicators of overwash deposits and barrier island sediment supply, Folly Beach, South Carolina. *Palaeogeography, Palaeoclimatology, Palaeoecology.* In press.

Hoffman, A. 1989a. Mass extinctions: the view of a sceptic. *Journal of the Geological Society of London* 146:21–35.

———. 1989b. Changing palaeontological views on mass extinction phenomena. In S. K. Donovan, ed., *Mass Extinctions: Processes and Evidence,* pp. 1–18. Columbia University Press, New York.

REFERENCES

Hoge, B. E. 1994. Wetland ecology and paleoecology: relationships between biogeochemistry and preservable taxa. *Current Topics in Wetland Biogeochemistry* 1:48–67.

Holland, S. M. 1995. The stratigraphic distribution of fossils. *Paleobiology* 21:92–109.

Holser, W. T., and M. Magaritz. 1987. Events near the Permian–Triassic boundary. *Modern Geology* 11:155–280.

——. 1992. Cretaceous/Tertiary and Permian/Triassic boundary events compared. *Geochimica et Cosmochimica Acta* 56:3297–3309.

Holser, W. T., M. Schidlowski, F. T. Mackenzie, and J. B. Maynard. 1988. Biogeochemical cycles of carbon and sulfur. In C. B. Gregor, R. M. Garrels, F. T. Mackenzie, and J. B. Maynard, eds., *Chemical Cycles in the Evolution of the Earth*, 105–173. Wiley, New York.

Holser, W. T., H.-P. Schönlaub, K. Boeckelmann, and M. Magaritz. 1991. The Permian-Triassic of the Gartnerkofel-1 Core (Carnic Alps, Austria): synthesis and conclusions. *Abhandlungen der Geologischen Bundesanstalt* 45: 213–232.

Hood, K. 1986. *GraphCor: Interactive Graphic Correlation for Microcomputers.* K. Hood, Houston, Tex.

Horn, H. S. 1975. Markovian properties of forest succession. In M. L. Cody and J. M. Diamond, eds., *Ecology and Evolution of Communities*, pp. 196–211. Harvard University Press, Harvard, Mass.

Horodyski, R. J., and L. P. Knauth. 1994. Life on land in the Precambrian. *Science* 263:494–498.

House, M. R. 1985. Correlation of mid-Palaeozoic ammonoid evolutionary events with global sedimentary perturbations. *Nature* 313:17–22.

——. 1989. Ammonoid extinction events. *Philosophical Transactions of the Royal Society of London* B325:307–326.

Hsü, K. J. 1986. Environmental changes in times of biotic crisis. In D. M. Raup and D. Jablonski, eds., *Patterns and Processes in the History of Life*, pp. 297–312. Springer-Verlag, Berlin.

Hudson, J. H., K. J. Hanson, R. B. Halley, and J. L. Kindinger. 1994. Environmental implications of growth rate changes in *Montastrea annularis:* Biscayne Bay Park, Florida. *Bulletin of Marine Science* 54:647–669.

Hut, P., W. Alvarez, W. P. Elder, T. A. Hansen, E. G. Kauffman, G. Keller, E. M. Shoemaker, and P. R. Weissman. 1987. Comet showers as a cause of mass extinctions. *Nature* 329:118–126.

Hyndman, R. D. 1995. Giant earthquakes of the Pacific northwest. *Scientific American* 273 (December):68–75.

Imbrie, J., and K. P. Imbrie. 1979. *Ice Ages: Solving the Mystery.* Harvard University Press, Cambridge, Mass.

Ingall, E. D., R. M. Bustin, and P. Van Cappellen. 1993. Influence of water column anoxia on the burial and preservation of carbon and phosphorus in marine shales. *Geochimica et Cosmochimica Acta* 57:303–316.

Ingram, B. L., R. Coccioni, A. Montanari, and F. M. Richter. 1994. Strontium isotopic composition of mid-Cretaceous seawater. *Science* 264:546–550.

Iturralde-Vinent, M. A. 1995. *Sedimentary Geology of Western Cuba*. Guide for post-congress field trip of the First SEPM Congress on Sedimentary Geology, St. Petersburg, Fla., August 13–16.

Jablonski, D. J. 1985. Marine regressions and mass extinctions: A test using the modern biota. In J. W. Valentine, ed., *Phanerozoic Diversity Patterns: Profiles in Macroevolution*, pp. 335–354. Princeton University Press, Princeton, N.J.

——. 1986a. Causes and consequences of mass extinctions: a comparative approach. In D. K. Elliott, ed., *Dynamics of Extinction*, pp. 183–229. Wiley, New York.

——. 1986b. Background and mass extinctions: the alternation of macroevolutionary regimes. *Science* 231:129–133.

——. 1989. The biology of mass extinction: a paleontological view. *Philosophical Transactions of the Royal Society of London* B325:357–368.

Jablonski, D. J., and D. J. Bottjer. 1989. Onshore–offshore trends in marine invertebrate evolution. In R. M. Ross and W. D. Allmon, eds., *Biotic and Abiotic Factors in Evolution*. University of Chicago Press, Chicago.

Jackson, J. B. C. 1988. Does ecology matter? *Paleobiology* 14:307–312.

——. 1991. Adaptation and diversity of reef corals. *BioScience* 41:475–482.

——. 1992. Pleistocene perspectives on coral reef community structure. *American Zoologist* 32:719–731.

——. 1994. Community unity? *Science* 264:1412–1413.

Jafar, S. A. 1983. Significance of Late Triassic calcareous nannoplankton from Austria and southern Germany. *Neue Jahrbuch für Geologie und Paläontologie* 166:218–259.

James, N. P. 1983. Reef. In P. Scholle, D. G. Bebout, and C. H. Moore, eds., *Carbonate Depositional Environments*, pp. 346–440. American Association of Petroleum Geologists, Tulsa, Okla.

Joachimski, M. M., and W. Buggisch. 1993. Anoxic events in the late Frasnian: causes of the Frasnian-Famennian faunal crisis? *Geology* 21:675–678.

John, S. J. 1995. Microfaunal assemblages: their use in interpreting depositional environments within a transgressive valley-fill sequence. *Geological Society of America Abstracts with Program* 27:28.

Johnson, J. G. 1974. Extinction of perched faunas. *Geology* 2:479–482.

Johnson, K. G., A. F. Budd, and T. A. Stemann. 1995. Extinction selectivity and ecology of Neogene Caribbean reef corals. *Paleobiology* 21:52–73.

Johnson, R. G. 1960. Models and methods for analysis of the mode of formation of fossil assemblages. *Geological Society of America Bulletin* 71:1075–1085.

Johnston, C. A. 1995. Effects of animals on landscape pattern. In L. Hansson, L. Fahrig, and G. Merriam, eds., *Mosaic Landscapes and Ecological Processes*, pp. 57–80. Chapman and Hall, London.

REFERENCES

Jones, C. E., H. C. Jenkyns, A. L. Coe, and S. P. Hesselbo. 1994. Strontium isotopic variations in Jurassic and Cretaceous seawater. *Geochimica et Cosmochimica Acta* 58:3061–3074.

Jones, D. J. 1956. *Introduction to Microfossils.* Hafner Publishing, New York (reprinted 1969).

Karlson, R. H., and L. E. Hurd. 1993. Disturbance, coral reef communities, and changing ecological paradigms. *Coral Reefs* 12:117–125.

Kauffman, E. G. 1994. Comparing modern and ancient biodiversity crises: the past as a key to the future. *Eos* (Supplement) 75:80.

Kauffman, E. G., and J. A. Fagerstrom. 1993. The Phanerozoic evolution of reef diversity. In R. E. Ricklefs and D. Schluter, eds., *Species Diversity in Ecological Communities,* pp. 315–329. University of Chicago Press, Chicago.

Kauffman, S. A. 1993. *The Origins of Order.* Oxford University Press, Oxford, England.

——. 1995. *At Home in the Universe.* Oxford University Press, Oxford, England.

Kaufman, A. J., S. B. Jacobsen, and A. H. Knoll. 1993. The Vendian record of Sr and C isotopic variations in seawater: implications for tectonics and paleoclimate. *Earth and Planetary Science Letters* 120:409–430.

Keith, M. L. 1982. Violent volcanism, stagnant oceans and some inferences regarding petroleum, strata-bound ores and mass extinctions. *Geochimica et Cosmochimica Acta* 46:2621–2637.

Keller, C. K., and B. D. Wood. 1993. Possibility of chemical weathering before the advent of vascular plants. *Nature* 364:223–225.

Kellert, S. H. 1993. *In the Wake of Chaos.* University of Chicago Press, Chicago.

Kelley, P. H., and T. A. Hansen. 1993. Evolution of the naticid gastropod predator–prey system: an evaluation of the hypothesis of escalation. *Palaios* 8:358–375.

Kerr, R. A. 1991. The stately cycles of ancient climate. *Science* 252:1254–1255.

——. 1994. Who profits from ecological disaster? *Science* 266:28–30.

——. 1995a. Studies say—tentatively—that greenhouse warming is here. *Science* 268:1567–1568.

——. 1995b. Scientists see greenhouse, semiofficially. *Science* 269:1667.

——. 1995c. Chesapeake Bay impact crater confirmed. *Science* 269:1672.

——. 1995d. A volcanic crisis for ancient life? *Science* 270:27–28.

——. 1996. Impact craters all in a row? *Science* 272:33.

Kidwell, S. M. 1986. Models for fossil concentrations: paleobiologic implications. *Paleobiology* 12:6–24.

——. 1988. Taphonomic comparison of passive and active continental margins: Neogene shell beds of the Atlantic coastal plain and northern Gulf of California. *Palaeogeography, Palaeoclimatology, Palaeoecology* 63:201–223.

——. 1989. Stratigraphic condensation of marine transgressive records: origin of major shell deposits in the Miocene of Maryland. *Journal of Geology* 97:1–24.

——. 1990. Phanerozoic evolution of macroinvertebrate shell accumulations: preliminary data from the Jurassic of Britain. In W. Miller, ed., *Paleocommunity Temporal Dynamics: The Long-Term Development of Multispecies Assemblies.* Paleontological Society, Special Publication No. 5:309–327

——. 1991. The stratigraphy of shell concentrations. In P. A. Allison and D. E. G. Briggs, eds., *Taphonomy: Releasing the Data Locked in the Fossil Record,* pp. 212–290. Plenum, New York.

——. 1993a. Time-averaging and temporal resolution in Recent marine shelly faunas. In S. M. Kidwell and A. K. Behrensmeyer, eds., *Taphonomic Approaches to Time Resolution in Fossil Assemblages.* Paleontological Society Short Courses in Paleontology No. 6:9–33.

——. 1993b. Taphonomic expressions of sedimentary hiatuses: field observations on bioclastic concentrations and sequence anatomy in low, moderate and high subsidence settings. *Geologische Rundschau* 82:189–202.

Kidwell, S. M., and A. K. Behrensmeyer. 1993. Summary: estimates of time-averaging. In S. M. Kidwell and A. K. Behrensmeyer, eds., *Taphonomic Approaches to Time Resolution in Fossil Assemblages.* Paleontological Society Short Courses in Paleontology No. 6:301–302.

Kidwell, S. M., and D. W. J. Bosence. 1991. Taphonomy and time-averaging of marine shelly faunas. In P. A. Allison and D. E. G. Briggs, eds., *Taphonomy: Releasing the Data Locked in the Fossil Record,* pp. 116–209. Plenum, New York.

Kidwell, S. M., and P. J. Brenchley. 1994. Patterns of bioclastic accumulation through the Phanerozoic: changes in input or destruction? *Geology* 22:1139–1143.

Kilham, P., and S. S. Kilham. 1980. The evolutionary ecology of phytoplankton. In I. Morris, ed., *The Physiological Ecology of Phytoplankton,* pp. 571–597. University of California Press, Berkeley.

Knoll, A. H. 1989. Evolution and extinction in the marine realm: some constraints imposed by phytoplankton. *Philosophical Transactions of the Royal Society of London* B325:279–290.

——. 1996. Daughters of time. *Paleobiology* 22:1–7.

Knoll, A. H., R. K. Bambach, D. E. Canfield, and J. P. Grotzinger. 1996. Comparative Earth history and Late Permian mass extinction. *Science* 273:452–457.

Knoll, A. H., and G. W. Rothwell. 1981. Paleobotany: perspectives in 1980. *Paleobiology* 7:7–35.

Knoll, A. H., and K. Swett. 1987. Micropaleontology across the Precambrian–Cambrian boundary in Spitsbergen. *Journal of Paleontology* 61:898–926.

Knoll, M. A., and W. C. James. 1987. Effect of the advent and diversification of vascular land plants on mineral weathering through geologic time. *Geology* 15:1099–1102.

Knowlton, N. 1992. Thresholds and multiple stable states in coral reef community dynamics. *American Zoologist* 32:674–682.

Knowlton, N., Lang, J. C., and B. D. Keller. 1990. Case study of natural population collapse: post-hurricane predation on Jamaican staghorn corals. *Smithsonian Contributions to Marine Sciences* 31:1–125.

Koch, C. F. 1987. Prediction of sample size effects on the measured temporal and geographic distribution patterns of species. *Paleobiology* 13:100–107.

Koeberl, C., C. W. Poag, W. U. Reimold, and D. Brandt. 1996. Impact origin of the Chesapeake Bay structure and the source of the North American tektites. *Science* 271:1263–1266.

Kotler, E., R. E. Martin, and W. D. Liddell. 1992. Experimental analysis of abrasion and dissolution resistance of modern reef-dwelling foraminifera: implications for the preservation of biogenic carbonate. *Palaios* 7:244–276.

Kowalewski, M. 1996. Time-averaging, overcompleteness, and the geological record. *Journal of Geology* 104:317–326.

Kramm, U., and K. H. Wedepohl. 1991. The isotopic composition of strontium and sulfur in seawater of Late Permian (Zechstein) age. *Chemical Geology* 90:253–262.

Krishnaswami, S., L. K. Benninger, R. C. Aller, and K. L. Vondamm. 1980. Atmospherically derived radionuclides as tracers of sediment mixing and accumulation in near shore marine and lake sediments: evidence from ^{7}Be, ^{210}Pb, and ^{239,240}Pu. *Earth and Planetary Science Letters* 47:307–318.

Krumenaker, L. 1994. In ancient climate, orbital chaos? *Science* 263:323.

Kump, L. R. 1989. Alternative modeling approaches to the geochemical cycles of carbon, sulfur, and strontium isotopes. *American Journal of Science* 289:390–410.

———. 1991. Interpreting carbon-isotope excursions: Strangelove oceans. *Geology* 19:299–302.

Kump, L. R., and F. T. Mackenzie. 1996. Regulation of atmospheric O_2: feedback in the microbial feedbag. *Science* 271:459–460.

Lal, D., and B. Peters. 1967. Cosmic-ray produced radioactivity on the earth. *Handbuch der Physik* 46:551–612.

Lande, R. 1994. Risk of population extinction from fixation of new deleterious mutations. *Evolution* 48:1460–1469.

Larson, D. W., and D.C. Rhoads. 1983. The evolution of infaunal communities and sedimentary fabrics. In M. J. S. Tevesz, and P. L. McCall, eds., *Biotic Interactions in Recent and Fossil Benthic Communities*, pp. 627–648. Plenum, New York.

Larson, R. L., and P. Olson. 1991. Mantle plumes control magnetic reversal frequency. *Earth and Planetary Science Letters* 107:437–447.

Laskar, J. 1989. A numerical experiment on the chaotic behavior of the solar system. *Nature* 338:237–238.

Levinton, J. S. 1996. Trophic group and the end-Cretaceous extinction: did deposit feeders have it made in the shade? *Paleobiology* 22:104–112.

Licari, G. R. 1978. Biogeology of the late Pre-Phanerozoic Beck Spring Dolomite of eastern California. *Journal of Paleontology* 52:767–792.

Lieberman, B. S., C. E. Brett, and N. Eldredge. 1995. A study of stasis and change in two species lineages from the Middle Devonian of New York state. *Paleobiology* 21:15–27.

Lin, S., and J. W. Morse. 1991. Sulfate reduction and iron sulfide mineral formation in Gulf of Mexico anoxic sediments. *American Journal of Science* 291:55–89.

Linke, P. and G. F. Lutze. 1993. Microhabitat preferences of benthic foraminifera—a static or a dynamic adaptation to optimize food acquisition? *Marine Micropaleontology* 20:215–234.

Lipps, J. H. 1970. Plankton evolution. *Evolution* 24:1–22.

Lipps, J. H., and K. McCartney. 1993. Chrysophytes. In J. H. Lipps, ed., *Fossil Prokaryotes and Protists*, pp. 141–154. Blackwell Scientific, Boston.

Logan, G. A., J. M. Hayes, G. B. Hieshima, and R. E. Summons. 1995. Terminal Proterozoic reorganization of biogeochemical cycles. *Nature* 376:53–56.

Loper, D. E., K. McCartney, and G. Buzyna. 1988. A model of correlated periodicity in magnetic-field reversals, climate, and mass extinctions. *Journal of Geology* 96:1–15.

Lord, A. R., and G. B. Hamilton. 1982. Paleozoic calcareous nannofossils. In A. R. Lord, ed., *A Stratigraphical Index of Calcareous Nannofossils*, p. 16. Halsted Press, New York.

Lord, C. J., and T. M. Church. 1983. The geochemistry of salt marshes: sedimentary ion diffusion, sulfate reduction, and pyritization. *Geochimica et Cosmochimica Acta* 47:1381–1391.

Loubere, P. 1989. Bioturbation and sedimentation rate control of benthic microfossil taxon abundances in surface sediments: a theoretical approach to the analysis of species microhabitats. *Marine Micropaleontology* 14:317–325.

——. 1991. Deep-sea benthic foraminiferal assemblage response to a surface ocean productivity gradient: a test. *Paleoceanography* 6:193–204.

Loubere, P., and A. Gary. 1990. Taphonomic process and species microhabitats in the living to fossil assemblage transition of deeper water benthic foraminifera. *Palaios* 5:375–381.

Loubere, P., A. Gary, and M. Lagoe. 1993a. Generation of the benthic foraminiferal assemblage: theory and preliminary data. *Marine Micropaleontology* 20:165–181.

——. 1993b. Sea-bed biogeochemistry and benthic foraminiferal zonation on the slope of the northwest Gulf of Mexico. *Palaios* 8:439–449.

Loubere, P., P. Meyers, and A. Gary. 1995. Benthic foraminiferal microhabitat selection, carbon isotope values, and association with larger animals: a test with *Uvigerina peregrina*. *Journal of Foraminiferal Research* 25:83–95.

Luther, G. W., T. M. Church, J. R. Scudlark, and M. Cosman. 1986. Inorganic and organic sulfur cycling in salt-marsh pore waters. *Science* 232:746–749.

Luther, G. W., T. G. Ferdelman, J. E. Kostka, E. J. Tsamakis, and T. M. Church. 1991. Temporal and spatial variability of reduced sulfur species $(FeS_2, S_2O_3^{2-})$ and porewater parameters in salt marsh sediments. *Biogeochemistry* 14:57–88.

REFERENCES

Lyman, R. L. 1994. *Vertebrate Taphonomy.* Cambridge University Press, Cambridge, England.

MacArthur, R. H., and E. O. Wilson. 1967. *The Theory of Island Biogeography.* Princeton University Press, Princeton, N.J.

Mackenzie, F. T., and C. Agegian. 1989. Biomineralization and tentative links to plate tectonics. In R. E. Crick, ed., *Origin, Evolution, and Modern Aspects of Biomineralization in Plants and Animals,* pp. 11–27. Plenum Press, New York.

Mackenzie, F. T., and J. W. Morse. 1992. Sedimentary carbonates through Phanerozoic time. *Geochimica et Cosmochimica Acta* 56:3281–3295.

Mackenzie, F. T., and J. D. Pigott. 1981. Tectonic controls of Phanerozoic sedimentary rock cycling. *Journal of the Geological Society of London* 138: 183–196.

MacLeod, K. G., and B. T. Huber. 1996. Strontium isotopic evidence for extensive reworking in sediments spanning the Cretaceous–Tertiary boundary at ODP site 738. *Geology* 24:463–466.

MacLeod, N. 1991. Punctuated anagenesis and the importance of stratigraphy to paleobiology. *Paleobiology* 17:167–188.

——. 1995. Graphic correlation of new Cretaceous/Tertiary (K/T) boundary sections/cores from Denmark, Alabama, Mexico, and the southern Indian Ocean: implications for a global sediment accumulation model. In K. O. Mann, H. R. Lane, and J. A. Stein, eds., *Graphic Correlation and the Composite Standard Approach.* Society for Sedimentary Geology Special Publication No. 53, Tulsa, Okla.

MacLeod, N., and G. Keller. 1991. How complete are Cretaceous/Tertiary boundary sections? A chronostratigraphic estimate based on graphic correlation. *Geological Society of America Bulletin* 103:1439–1457.

Magaritz, M., H. T. Holser, and J. L. Kirschvink. 1986. Carbon-isotope events across the Precambrian–Cambrian boundary on the Siberian platform. *Nature* 320:258–259.

Magaritz, M., R. V. Kristhnamurthy, and W. T. Holser. 1992. Parallel trends in organic and inorganic carbon isotopes across the Permian/Triassic boundary. *American Journal of Science* 292:727–739.

Maliva, R. G., A. H. Knoll, and R. Siever. 1989. Secular change in chert distribution: a reflection of evolving biological participation in the silica cycle. *Palaios* 4:519–532.

Małkowski, K., M. Gruszczynski, A. Hoffman, and S. Halas. 1989. Oceanic stable isotope composition and a scenario for the Permo-Triassic crisis. *Historical Biology* 2:289–309.

Mandelbrot, B. 1983. *The Fractal Geometry of Nature.* W. H. Freeman, New York.

Mann, K. O., H. R. Lane, and J. A. Stein, eds. 1995. *Graphic Correlation and the Composite Standard Approach to Biostratigraphy.* Society for Sedimentary Geology Special Publication No. 53, Tulsa, Okla.

Margalef, R. 1968. *Perspectives in Ecological Theory: University of Chicago Press*, Chicago.

———. 1971. The pelagic ecosystem of the Caribbean Sea. *Symposium on Investigations and Resources of the Caribbean Sea and Adjacent Regions*, pp. 483–498. UNESCO, Paris.

Margulis, L. E. 1970. *Origin of Eukaryotic Cells*. Yale University Press, New Haven, Conn.

Margulis, L. E., E. S. Barghoorn, D. Ashendorf, S. Banerjee, D. Chase, S. Francis, S. Giovannoni, and S. Stolz. 1980. The microbial community in the layered sediments at Laguna Figueroa, Baja, California, Mexico: does it have Precambrian analogues? *Precambrian Research* 11:92–123.

Marshall, C. R. 1990. Confidence intervals on stratigraphic ranges. *Paleobiology* 16:1–10.

———. 1995. Distinguishing between sudden and gradual extinctions in the fossil record: predicting the position of the Cretaceous–Tertiary iridium anomaly using the ammonite fossil record on Seymour Island, Antarctica. *Geology* 23:731–734.

Martin, R. E. 1991. Beyond biostratigraphy: micropaleontology in transition? *Palaios* 6:437–438.

———. 1993. Time and taphonomy: actualistic evidence for time-averaging of benthic foraminiferal assemblages. In S. M. Kidwell and A. K. Behrensmeyer, eds., *Taphonomic Approaches to Time Resolution in Fossil Assemblages*, pp. 34–56. Paleontological Society Short Courses in Paleontology No.6.

———. 1995a. Cyclic and secular variation in microfossil biomineralization: clues to the biogeochemical evolution of Phanerozoic oceans. *Global and Planetary Change* 11:1–23.

———. 1995b. The once and future profession of micropaleontology. *Journal of Foraminiferal Research* 25:372–373.

———. 1996a. Secular increase in nutrient levels through the Phanerozoic: implications for productivity, biomass, and diversity of the marine biosphere. *Palaios* 11:209–219.

———. 1996b. Catastrophic fluctuations in nutrient levels as an agent of mass extinction: upward scaling of ecological processes? In M. L. McKinney and J. A. Drake, eds., *Biodiversity Dynamics: Origination and Extinction of Populations, Species, Communities, and Higher Taxa*. Columbia University Press, New York. In press.

Martin, R. E., and R. R. Fletcher. 1995. Graphic correlation of Plio-Pleistocene sequence boundaries, Gulf of Mexico: oxygen isotopes, ice volume, and sea level. In K. O. Mann, H. R. Lane, and J. A. Stein, eds., *Graphic Correlation and the Composite Standard Approach*, pp. 235–248. Society for Sedimentary Geology Special Publication No. 53, Tulsa, Okla.

Martin, R. E., M. S. Harris, and W. D. Liddell. 1996a. Taphonomy and time-averaging of foraminiferal assemblages in Holocene tidal flat sediments, Bahia la Choya, Sonora, Mexico (northern Gulf of California). *Marine Micropaleontology* 26:187–206.

REFERENCES

Martin, R. E., G. W. Johnson, E. D. Neff, and D. E. Krantz. 1990. Quaternary planktonic foraminiferal assemblage zones of the northeast Gulf of Mexico, Colombia Basin (Caribbean Sea), and tropical Atlantic Ocean: graphic correlation of microfossil and oxygen isotope datums. *Paleoceanography* 5: 531–555.

Martin, R. E., and W. D. Liddell. 1988. Foraminiferal biofacies on a north coast fringing reef (1–75 m), Discovery Bay, Jamaica. *Palaios* 3:298–314.

——. 1991. The taphonomy of foraminifera in modern carbonate environments: implications for the formation of foraminiferal assemblages. In S. K. Donovan, ed., *The Processes of Fossilization*, 170–193. Belhaven, London.

Martin, R. E., E. D. Neff, G. W. Johnson, and D. E. Krantz. 1993. Biostratigraphic expression of Pleistocene sequence boundaries, Gulf of Mexico. *Palaios* 8:155–171.

Martin, R. E., J. F. Wehmiller, M. S. Harris, and W. D. Liddell. 1996b. Comparative taphonomy of foraminifera and bivalves in Holocene shallow-water carbonate and siliciclastic regimes: taphonomic grades and temporal resolution. *Paleobiology* 22:80–90.

Martin, R. E., and R. C. Wright. 1988. Information loss in the transition from life to death assemblages of foraminifera in back reef environments, Key Largo, Florida. *Journal of Paleontology* 62:399–410.

Maslin, M. A., N.J. Shackleton, and U. Pflaumann. 1995. Surface water temperature, salinity, and density changes in the northeast Atlantic during the last 45,000 years: Heinrich events, deep water formation, and climatic rebounds. *Paleoceanography* 10:527–544.

Mathewes, R. W., and J. J. Clague. 1994. Detection of large prehistoric earthquakes in the Pacific Northwest by microfossil analysis. *Science* 264: 688–691.

Matisoff, G. 1982. Mathematical models of bioturbation. In P. L. McCall and M. J. S. Tevesz, eds., *Animal-Sediment Relations*, pp. 289–330. Plenum Press, New York.

Maurer, B. A., and M. P. Nott. 1996. Geographic range fragmentation and the evolution of biological diversity. In M. L. McKinney and J. A. Drake, eds., *Biodiversity Dynamics: Origination and Extinction of Populations, Species, Communities, and Higher Taxa*. Columbia University Press, New York. In press.

May, R. M. 1973a. Time-delay versus stability in population models with two and three trophic levels. *Ecology* 54:315–325.

——. 1973b. *Stability and Complexity in Model Ecosystems*. Princeton University Press, Princeton, N.J.

——. 1976. Patterns in multi-species communities. In R. May, ed., *Theoretical Ecology: Principles and Applications*, pp. 142–162. Saunders, Philadelphia.

——. 1977. Thresholds and breakpoints in ecosystems with a multiplicity of stable states. *Nature* 269:471–477.

——. 1994. The effects of spatial scale on ecological questions and answers.

In P. J. Edwards, R. M. May, and N. R. Webb, eds., *Large-Scale Ecology and Conservation Biology*, pp. 1–17. Blackwell, London.

May, R. M., and G. F. Oster. 1976. Bifurcations and dynamic complexity in simple ecological models. *American Naturalist* 110:573–599.

Mayr, E. 1942. *Systematics and the Origin of Species.* Columbia University Press, New York.

McCave, I. N., B. Manighetti, and N. A. S. Beveridge. 1995. Circulation in the glacial North Atlantic inferred from grain-size measurements. *Nature* 374: 149–151.

McCorkle, D.C., L. D. Keigwin, B. H. Corliss, and S. R. Emerson. 1990. The influence of microhabitats on the carbon isotopic composition of deep-sea benthic foraminifera. *Paleoceanography* 5:161–185.

McGhee, G. R. 1996. *The Late Devonian Mass Extinction: The Frasnian–Famennian Crisis.* Columbia University Press, New York.

McGowan, J. A. 1971. Oceanic biogeography of the Pacific. In B. M. Funnell and W. R. Riedel, eds., *The Micropalaeontology of Oceans*, pp. 3–74. Cambridge University Press, Cambridge, England.

——. 1986. Beyond classical biostratigraphy. *Petroleum Exploration Society of Australia Bulletin* 9 (September):28–41.

McKinney, M. L. 1989. Periodic mass extinctions: product of biosphere growth dynamics? *Historical Biology* 2:273–287.

——. 1991. Completeness of the fossil record: an overview. In S. K. Donovan, ed., *The Processes of Fossilization*, pp. 66–83. Belhaven Press, London.

——. 1995. Extinction selectivity among lower taxa: gradational patterns and rarefaction error in extinction estimates. *Paleobiology* 21:300–313.

——. 1996. The biology of fossil abundance. *Revista Española de Paleontologia* 11:125–133.

McKinney, M. L., and W. D. Allmon. 1995. Metapopulations and disturbance: from patch dynamics to biodiversity dynamics. In D. H. Erwin and R. L. Anstey, eds., *New Approaches to Speciation in the Fossil Record.* Columbia University Press, New York.

McKinney, M. L., and D. Frederick. 1992. Extinction and population dynamics: new methods and evidence from Paleogene foraminifera. *Geology* 20: 343–346.

McKinney, M. L., J. L. Lockwood, and D. R. Frederick. 1996. Rare species and scale- dependence in ecosystem stasis. In L. C. Ivany and K. M. Schopf, eds., *New Perspectives on Faunal Stability in the Fossil Record. Palaeogeography, Palaeoclimatology, Palaeoecology* 127:191–207.

McMenamin, M. A. S., and D. L. S. McMenamin. 1990. *The Emergence of Animals: The Cambrian Breakthrough.* Columbia University Press, New York.

——. 1994. *Hypersea: Life on Land.* Columbia University Press, New York.

McRoberts, C. A., and C. R. Newton. 1995. Selective extinction among end-Triassic European bivalves. *Geology* 23:102–104.

Meldahl, K. H. 1987. Sedimentologic and taphonomic implications of biogenic stratification. *Palaios* 2:350–358.

239

REFERENCES

———. 1990. Sampling, species abundance, and the stratigraphic signature of mass extinction: a test using Holocene tidal flat molluscs. *Geology* 18:890–893.

Mereschkowsky, C. 1910. Theorie der zwei Plasmaarten als Grundlage der Symbiogenesis einer neuen Lehre von der Entstehung der Organismen. Cited by F. J. R. Taylor. 1987. An overview of the status of evolutionary cell symbiosis theories. *Annals of the New York Academy of Sciences* 503:1–16.

Miall, A. D. 1990. *Principles of Sedimentary Basin Analysis.* Springer-Verlag, New York.

Mii, H.-S., E. L. Grossman, and T. E. Yancey. 1997. Stable carbon and oxygen isotope shifts in Permian seas of West Spitsbergen: Global change or diagenetic artifact? *Geology* 25:227–230.

Miller, F. X. 1977. The graphic correlation method in biostratigraphy. In E. G. Kauffman and J. E. Hazel, eds., *Concepts and Methods of Biostratigraphy,* pp. 165–186. Dowden, Hutchinson, and Ross, Stroudsburg, Pa.

Miller, K. G., M. D. Feigenson, D. V. Kent, and R. K. Olsson. 1988. Upper Eocene to Oligocene isotope ($^{87}Sr/^{86}Sr$, $\delta^{18}O$, $\delta^{13}C$) standard section, DSDP Site 522. *Paleoceanography* 3:223–233.

Miller, W. 1986. Paleoecology of benthic community replacement. *Lethaia* 19:225–231.

———. 1991. Hierarchical concept of reef development. *Neues Jahrbuch für Geologie und Paläontologie Abhandlung* 182:21–35.

Mintz, L. W. 1981. *Historical Geology: The Science of a Dynamic Earth.* Merrill, Columbus, Ohio.

Monastersky, R. 1994. Staggering through the Ice Ages: what made the planet careen through climate extremes? *Science News* 146:74–76.

———. 1995. Pine forest thrives on high-CO_2 diet. *Science News* 148:101.

Moore, G. T., D. N. Hayashida, and C. A. Ross. 1993. Late Early Silurian (Wenlockian) general circulation model-generated upwelling, graptolitic black shales, and organic-rich source rocks: an accident of plate tectonics? *Geology* 21:17–20.

Moore, J. A. 1993. *Science as a Way of Knowing: The Foundations of Modern Biology.* Harvard University Press, Cambridge, Mass.

Moore, T. C., N. G. Pisias, and D. A. Dunn. 1982. Carbonate time series of the Quaternary and late Miocene sediments in the Pacific Ocean: a spectral comparison. *Marine Geology* 46:217–234.

Morris, P. J., and L. C. Ivany. 1994. Response to environmental change in light of ecological locking. *Eos* (Supplement) 75:80–81.

Morse, J. W., and F. T. Mackenzie. 1990. *Geochemistry of Sedimentary Carbonates.* Elsevier, Amsterdam.

Morse, J. W., Q. Wang, and Tsio, M. Y. 1997. Influences of temperature and Mg:Ca ratio on $CaCO_3$ precipitates from seawater. *Geology* 25:85–87.

Murray-Wallace, C. V., and A. P. Belperio. 1995. Identification of remanié fossils using amino acid racemisation. *Alcheringa* 18:219–227.

Nelson, A. R., and A. E. Jennings. 1993. Intertidal foraminifera and earthquake hazard assessment in the Cascadia subduction zone. *Geological Society of America Abstracts with Programs* 25:138.

Newell, N. D. 1967. Revolutions in the history of life. *Geological Society of America Special Paper* 89:63–91.

Nicol, D. 1962. The biotic development of some Niagaran reefs: an example of an ecological succession or sere. *Journal of Paleontology* 36:172–176.

Nicolis, G., and I. Prigogine. 1989. *Exploring Complexity: An Introduction.* W. H. Freeman, San Francisco.

Odum, E. P. 1969. The strategy of ecosystem development. *Science* 164: 262–270.

Officer, C. B. 1982. Mixing, sedimentation rates and age dating for sediment cores. *Marine Geology* 46:261–278.

Officer, C. B., and D. R. Lynch. 1983. Determination of mixing parameters from tracer distributions in deep-sea sediment cores. *Marine Geology* 52: 59–74.

Officer, C. B., and J. Page. 1996. *The Great Dinosaur Extinction Controversy,* Addison-Wesley, Reading, Mass.

Olson, E. C. 1980. Taphonomy: its history and role in community evolution. In A. K. Behrensmeyer and A. P. Hill, eds., *Fossils in the Making: Vertebrate Taphonomy and Paleoecology,* pp. 5–19. University of Chicago, Chicago.

O'Neill, R. V. 1988. Hierarchy theory and global change. In T. Rosswall, R. G. Woodmansee, and P. G. Risser, eds., *Scales and Global Change,* pp. 29–45. Wiley, New York.

O'Neill, R. V., D. L. DeAngelis, J. B. Waide, and T. F. H. Allen. 1986. *A Hierarchical Concept of Ecosystems.* Princeton University Press, Princeton. N.J.

O'Neill, R. V., A. R. Johnson, and A. W. King. 1989. A hierarchical framework for the analysis of scale. *Landscape Ecology* 3:193–205.

Oppo, D., and R. G. Fairbanks. 1987. Variability in the deep and intermediate water circulation of the Atlantic Ocean during the past 25,000 years: northern hemisphere modulation of the Southern Ocean. *Earth and Planetary Science Letters* 86:1–15.

Oppo, D., and S. J. Lehman. 1995. Suborbital timescale variability of North Atlantic Deep Water during the past 200,000 years. *Paleoceanography* 10: 901–910.

Orth, C. J. 1988. Geochemistry of the bio-event horizons. In S. K. Donovan, ed., *Mass Extinctions: Processes and Evidence,* pp. 37–72. Columbia University Press, New York.

Orth, C. J., J. S. Gilmore, L. R. Quintana, and P. M. Sheehan. 1986. Terminal Ordovician extinction: geochemical analysis of the Ordovician/Silurian boundary, Anticosti Island, Quebec. *Geology* 14:433–436.

Osborn, H. F. 1918. *The Origin and Evolution of Life.* Charles Scribner's Sons, New York.

Pandolfi, J. M. 1996. Limited membership in Pleistocene reef coral assemblages from the Huon Peninsula, Papua New Guinea: constancy during global change. *Paleobiology* 22:152–176.

REFERENCES

Parrish, J. T. 1982. Upwelling and petroleum source beds, with reference to Paleozoic. *American Association of Petroleum Geologists Bulletin* 66: 750–774.

———. 1987. Palaeo-upwelling and the distribution of organic-rich rocks. In J. Brooks and A. J. Fleet, eds., *Marine Petroleum Source Rocks*, pp. 199–205. Geological Society of London, London.

———. 1993. Climate of the supercontinent Pangea. *Journal of Geology* 101: 215–233.

Patterson, R. T., and Fowler, A. D. 1996. Evidence of self organization in planktic foraminiferal evolution: implications for interconnectedness of paleoecosystems. *Geology* 24:215–218.

Patterson, R. T., D. L. Ozarko, J.-P. Guilbault, and J. J. Clague. 1994. Distribution and preservation potential of marsh foraminiferal biofacies from the lower mainland and Vancouver Island, British Columbia. *Geological Society of America Abstracts with Programs* 26:530.

Peak, D., and M. Frame. 1994. *Chaos Under Control: The Art and Science of Complexity*. W. H. Freeman, New York.

Peterson, C. H. 1977. The paleoecological significance of undetected short-term temporal variability. *Journal of Paleontology* 51:976–981.

Peterson, I. 1993. *Newton's Clock: Chaos in the Solar System*. W. H. Freeman, San Francisco.

Phillips, F. J. 1986. A review of graphic correlation. *Computer Oriented Geological Society Contributions* 2:73–91.

Pimm, S. L. 1991. *The Balance of Nature?: Ecological Issues in the Conservation of Species and Communities*. University of Chicago Press, Chicago.

Pitrat, C. W. 1970. Phytoplankton and the late Paleozoic wave of extinction. *Palaeogeography, Palaeoclimatology, Palaeoecology* 8:49–55.

Pizzuto, J. E., and A. E. Schwendt. 1997. Mathematical modeling of autocompaction of a Holocene transgressive valley-fill deposit, Wolfe Glade, Delaware. *Geology* 25:57–60.

Plotnick, R. E. 1986. A fractal model for the distribution of stratigraphic hiatuses. *Journal of Geology* 94:885–890.

Plotnick, R. E., and M. L. McKinney. 1993. Ecosystem organization and ecosystem dynamics. *Palaios* 8:202–212.

Polis, G. A., and D. R. Strong. 1996. Food web complexity and community dynamics. *American Naturalist* 147:813–846.

Pool, R. 1989. Ecologists flirt with chaos. *Science* 243:310–313.

Popp, B. N., T. F. Anderson, and P. A. Sandberg. 1986. Brachiopods as indicators of original isotopic compositions in some Paleozoic limestones. *Geological Society of America Bulletin* 97:1262–1269.

Pospichal, J. J. 1994. Calcareous nannofossils at the K-T boundary, El Kef: no evidence for stepwise, gradual, or sequential extinctions. *Geology* 22:99–102.

Powell, E. N., H. Cummins, R. J. Stanton, and G. Staff. 1984. Estimation of the size of molluscan larval settlement using the death assemblage. *Estuarine and Coastal Shelf Science* 18:367–384.

Prothero, D. R. 1994. *The Eocene–Oligocene Transition*. Columbia University Press, New York.

Rahmstorf, S. 1995. Bifurcations of the Atlantic thermohaline circulation in response to changes in the hydrological cycle. *Nature* 378:145–150.

Railsback, L. B. 1993. Original mineralogy of Carboniferous worm tubes: evidence for changing marine chemistry and biomineralization. *Geology* 21:703–706.

Railsback, L. B., S. C. Ackerly, T. F. Anderson, J. L. and Cisne. 1990. Palaeontological and isotope evidence for warm saline deep waters in Ordovician seas. *Nature* 343:156–159.

Railsback, L. B., and T. F. Anderson. 1987. Control of Triassic seawater chemistry and temperature on the evolution of post-Paleozoic aragonite-secreting faunas. *Geology* 15:1002–1005.

Raiswell, R., and R. A. Berner. 1986. Pyrite and organic matter in Phanerozoic normal marine shales. *Geochimica et Cosmochimica Acta* 50:1967–1976.

Rampino, M. R. 1993. Asteroid impacts, mass extinction events, and flood basalt eruptions: an external driver. Stratigraphic Record of Global Change, SEPM Meeting, Pennsylvania State University, August 8–12, *Abstracts with Program*, pp. 57–58.

Rampino, M. R., and K. Caldeira. 1993. Major episodes of geologic change: correlations, time structure and possible causes. *Earth and Planetary Sciences* 114:215–227

Rampino, M. R., and B. M. Haggerty. 1995. Mass extinctions and periodicity. *Science* 269:617–619.

Rampino, M. R., S. Self, and R. B. Stothers. 1988. Volcanic winters. *Annual Review of Earth and Planetary Sciences* 16:73–100.

Rau, G., M. A. Arthur, and W. E. Dean. 1987. $^{15}N/^{14}N$ variations in Cretaceous Atlantic sedimentary sequences: implications for past changes in marine nitrogen biogeochemistry. *Earth and Planetary Science Letters* 82:269–279.

Raup, D. M. 1989. The case for extraterrestrial causes of extinction. *Philosophical Transactions of the Royal Society of London* 325B:421–435.

——. 1991. *Extinction: Bad Genes or Bad Luck?* W. W. Norton, New York.

——. 1994. The role of extinction in evolution. *Proceedings of the National Academy of Sciences* 91:6758–6763.

Raup, D. M., and J. J. Sepkoski. 1982. Mass extinctions in the marine fossil record. *Science* 215:1501–1503.

Raymo, M. E. 1991. Geochemical evidence supporting T. C. Chamberlin's theory of glaciation. *Geology* 19:344–347.

——. 1994. The Himalayas, organic carbon burial, and climate in the Miocene. *Paleoceanography* 9:399–404.

Raymo, M. E., W. F. Ruddiman, and P. N. Froelich. 1988. Influence of late Cenozoic mountain building on ocean geochemical cycles. *Geology* 16:649–653.

Reice, S. R. 1994. Nonequilibrium determinants of biological community structure. *American Scientist* 82:424–435.

REFERENCES

Renard, M. 1986. Pelagic carbonate chemostratigraphy (Sr, Mg, ^{18}O, ^{13}C). *Marine Micropaleontology* 10:117–164.

Renne, P. R., Z. Zichao, M. A. Richards, M. T. Black, and A. R. Basu. 1995. Synchrony and causal relations between Permian–Triassic boundary crises and Siberian flood volcanism. *Science* 269:1413–1416.

Rhoads, D.C., and D. J. Stanley. 1965. Biogenic graded bedding. *Journal of Sedimentary Petrology* 35:956–963.

Rhodes, M. C., and C. W. Thayer. 1991. Mass extinctions: ecological selectivity and primary production. *Geology* 19:877–880.

Rhodes, M. C., and R. J. Thompson. 1993. Comparative physiology of suspension-feeding in living brachiopods and bivalves: evolutionary implications. *Paleobiology* 19:322–334.

Richmond, R. H. 1989. Competency and dispersal potential of planula larvae of a spawning versus a brooding coral. *Proceedings 6th International Coral Reef Symposium* 2:827–831.

———. 1990. The effects of El Niño/Southern Oscillation on the dispersal of corals and other marine organisms. In P. W. Glynn, ed., *Global Ecological Consequences of the 1982–83 El Niño-Southern Oscillation*, pp. 127–140. Elsevier, New York.

Richmond, R. H., and C. L. Hunter. 1990. Reproduction and recruitment of corals: comparisons among the Caribbean, the tropical Pacific, and the Red Sea. *Marine Ecology Progress Series* 60:185–203.

Richter, F. M., D. B. Rowley, and D. J. DePaolo. 1992. Sr isotope evolution of seawater: the role of tectonics. *Earth and Planetary Science Letters* 109:11–23.

Riding, R. 1993. Phanerozoic patterns of marine CaCO$_3$ precipitation. *Naturwissenschaften* 80:513–515.

Risk, M. J., R. D. Venter, S. G. Pemberton, and D. E. Buckley. 1978. Computer simulation and sedimentological implications of burrowing by *Axius serratus*. *Canadian Journal of Earth Sciences* 15:1370–1374.

Robbins, J. A. 1978. Geochemical and geophysical applications of radioactive lead. In J. O. Nriagu, ed., *The Biogeochemistry of Lead in the Environment*, pp. 285–393. Elsevier, Amsterdam.

———. 1982. Stratigraphic and dynamic effects of sediment reworking by Great Lakes zoobenthos. *Hydrobiologia* 92:611–622.

———. 1986. A model for particle-selective transport of tracers in sediments with conveyor belt deposit feeders. *Journal of Geophysical Research* 91: 8542–8558.

Robinson, J. M. 1990. Lignin, land plants, and fungi: biological evolution affecting Phanerozoic oxygen balance. *Geology* 15:607–610.

Ross, C. A. 1972. Paleobiological analysis of fusulinacean shell morphology. *Journal of Paleontology* 46:719–728.

Ruddiman, W. F., and L. K. Glover. 1972. Vertical mixing of ice-rafted volcanic ash in North Atlantic sediments. *Geological Society of America Bulletin* 83:2817–2836.

Ruddiman, W. F., G. Jones, T.-H. Peng, L. Glover, B. Glass, and P. Liebertz.

1980. Tests for size and shape dependency in deep-sea mixing. *Sedimentary Geology* 25:257–276.

Ruelle, D. *Chance and Chaos.* Princeton University Press, Princeton, N.J.

Runnels, C. N. 1995. Environmental degradation in ancient Greece. *Scientific American* 272 (March):96–99.

Ruttenberg, K. C., and R. A. Berner. 1993. Authigenic apatite formation and burial in sediments from non-upwelling, continental margin environments. *Geochimica et Cosmochimica Acta* 57:991–1007.

Sadler, P. M. 1981. Sediment accumulation rates and the completeness of stratigraphic sections. *Journal of Geology* 89:569–584.

Sadler, P. M., and D. J. Strauss. 1990. Estimation of completeness of stratigraphical sections using empirical data and theoretical models. *Journal of the Geological Society of London* 147:471–485.

Salmon, W. C. 1967. *The Foundations of Scientific Inference.* University of Pittsburgh Press, Pittsburgh.

Salthe, S. N. 1985. *Evolving Hierarchical Systems: Their Structure and Representation.* Columbia University Press, New York.

Sandberg, P. A. 1983. An oscillating trend in Phanerozoic non-skeletal carbonate mineralogy. *Nature* 305:19–22.

Sawlowicz, Z. 1993. Iridium and other platinum-group elements as geochemical markers in sedimentary environments. *Palaeogeography, Palaeoclimatology, Palaeoecology* 104:253–270.

Schidlowski, M. 1991. Quantitative evolution of global biomass through time: Biological and geochemical constraints. In S. H. Schneider and P. J. Boston, eds., *Scientists on Gaia,* pp. 211–222. MIT Press, Cambridge, Mass.

Schiffelbein, P. 1984. Effect of benthic mixing on the information content of deep-sea stratigraphic signals. *Nature* 311:651–653.

Schindel, D. E. 1980. Microstratigraphic sampling and the limits of paleontologic resolution. *Paleobiology* 6:408–426.

——. 1982. Resolution analysis: a new approach to the gaps in the fossil record. *Paleobiology* 8:340–353.

Schopf, J. W. 1996. Precambrian: the age of microscopic life. In J. E. Repetski, ed., *Sixth North American Paleontological Convention, Abstracts and Program,* p. 345. Paleontological Society Special Publication No. 8.

Schrödinger, E. 1944. *What Is Life?* Cambridge University Press, Cambridge, England.

Scott, D. B., and F. S. Medioli. 1980. Quantitative studies on marsh foraminiferal distributions in Nova Scotia: implications for sea level studies. Cushman Foundation for Foraminiferal Research Special Publication 17.

——. 1986. Foraminifera as sea-level indicators. In O. van de Plassche, ed., *Sea Level Research: A Manual for the Collection of Data,* pp. 435–455. Geo Books, Norwich, England (IGCP Projects 61 and 200).

Scott, P. J. B., M. J. Risk, and J. D. Carriquiry. 1988. El Niño, bioerosion and the survival of East Pacific Reefs. *Proceedings 6th International Coral Reef Symposium* 2:517–520.

REFERENCES

Sepkoski, J. J. 1986. Phanerozoic overview of mass extinction. In D. M. Raup and D. Jablonski, eds., *Patterns and Processes in the History of Life*, pp. 277–295. Springer-Verlag, Berlin.

——. 1989. Periodicity in extinction and the problem of catastrophism in the history of life. *Journal of the Geological Society of London* 146:7–19.

——. 1992. Phylogenetic and ecologic patterns in the Phanerozoic history of marine diversity. In N. Eldredge, ed., *Systematics, Ecology, and the Biodiversity Crisis*, pp. 77–100. Columbia University Press, New York.

Sepkoski, J. J., R. K. Bambach, and M. L. Droser. 1991. Secular changes in Phanerozoic event bedding and the biological overprint. In G. Einsele, W. Ricken, and A. Seilacher, eds., *Cycles and Events in Stratigraphy*, pp. 298–312. Springer-Verlag, Berlin.

Sepkoski, J. J., and A. I. Miller. 1985. Evolutionary faunas and the distribution of Paleozoic marine communities in space and time. In J. W. Valentine, ed., *Phanerozoic Diversity Patterns: Profiles in Macroevolution*, pp. 153–190. Princeton University Press, Princeton, N.J.

Shackleton, N.J. 1986. Paleogene stable isotope events. *Palaeogeography, Palaeoclimatology, Palaeoecology* 57:91–102.

Shackleton, N.J., and N. D. Opdyke. 1973. Oxygen isotope and paleomagnetic stratigraphy of equatorial Pacific core V28–238: oxygen isotope temperatures and ice volumes on a 10^5 and 10^6 year scale. *Quaternary Research* 3:39–55.

Sharma, P., L. R. Gardner, W. S. Moore, and M. S. Bollinger. 1987. Sedimentation and bioturbation in a salt marsh as revealed by ^{210}Pb, ^{137}Cs, and ^7Be studies. *Limnology and Oceanography* 32:313–326.

Shaw, A. B. 1964. *Time in Stratigraphy.* McGraw-Hill, New York.

Sheehan, P. 1985. Reefs are not so different: they follow the evolutionary pattern of level-bottom communities. *Geology* 13:46–49.

Sheehan, P., and T. A. Hansen. 1986. Detritus feeding as a buffer to extinction at the end of the Cretaceous. *Geology* 14:868–870.

Sheldon, R. P. 1980. Episodicity of phosphate deposition and deep ocean circulation: a hypothesis. In Y. K. Bentor, ed., *Marine Phosphorites: Geochemistry, Occurrence, Genesis*, pp. 239–247. Society of Economic Paleontologists and Mineralogists, Tulsa, Okla., Special Publication No. 29.

Siesser, W. R. 1993. Calcareous nannoplankton. In J. H. Lipps, ed., *Fossil Prokaryotes and Protists*, pp. 169–201. Blackwell Scientific, Boston.

Signor, P. W. 1992. Taxonomic diversity and faunal turnover in the Early Cambrian: did the most severe mass extinction of the Phanerozoic occur in the Botomian Stage? In S. Lidgard and P. R. Crane, eds., *Fifth North American Paleontological Convention, Abstracts and Program*, p. 272. Paleontological Society Special Publication No. 6.

——. 1994. Biodiversity in geological time. *American Zoologist* 34:23–32.

Signor, P. W., and C. W. Brett. 1984. The mid-Paleozoic precursor to the Mesozoic marine revolution. *Paleobiology* 10:229–245.

Signor, P. W., and J. H. Lipps. 1982. Sampling bias, gradual extinction patterns and catastrophes in the fossil record. In L. T. Silver and P. H. Schulz, eds.

Geological Implications of Impacts of Large Asteroids and Comets on the Earth, pp. 291–296. Geological Society of America Special Paper 190.

Signor, P. W., and G. J. Vermeij. 1994. The plankton and the benthos: origins and early history of an evolving relationship. *Paleobiology* 20:297–319.

Singer, A. J., and A. Shemesh. 1995. Climatically linked carbon isotope variation during the past 430,000 years in Southern Ocean sediments. *Paleoceanography* 10:171–177.

Sloan, D. 1995. Use of foraminiferal biostratigraphy in mitigating pollution and seismic problems, San Francisco, California. *Journal of Foraminiferal Research* 25:260–266.

Slobodkin, L. B. 1996. Islands of peril and pleasure. *Nature* 381:205–206.

Sloss, L. L. 1963. Sequences in the cratonic interior of North America. *Geological Society of America Bulletin* 74:93–114.

Smalley, P. C., A. C. Higgins, R. J. Howarth, H. Nicholson, C. E. Jones, N. M. H. Swinburne, and J. Bessa. 1994. Seawater Sr isotope variations through time: a procedure for constructing a reference curve to date and correlate marine sedimentary rocks. *Geology* 22:431–434.

Smith, A. B., and D. T. J. Littlewood. 1994. Paleontological data and molecular phylogenetic analysis. *Paleobiology* 20:259–273.

Smith, A. J. 1995. Ostracodes as indicators of climatic processes and anthropogenic disturbances in aquatic environments. *Geological Society of America Abstracts with Program* 27:A27

Smith, S. V., W. J. Kimmerer, E. A. Laws, R. E. Brock, and T. W. Walsh. 1981. Kaneohe Bay sewage diversion experiment: perspectives on ecosystem responses to nutritional perturbation. *Pacific Science* 35:279–395.

Speyer, S. E., and C. E. Brett. 1988. Taphofacies models for epeiric seas. *Palaeogeography, Palaeoclimatology, Palaeoecology* 63:225–262.

Staff, G. M., R. J. Stanton, E. N. Powell, and H. Cummins. 1986. Time-averaging, taphonomy, and their impact on paleocommunity reconstruction: death assemblages in Texas bays. *Geological Society of America Bulletin* 97:428–443.

Stanley, S. M. 1979. *Macroevolution: Pattern and Process.* W. H. Freeman, San Francisco.

——. 1984. Changes in sea level. In H. D. Holland and A. F. Trendall, eds., *Patterns of Change in Earth Evolution*, pp. 103–121. Springer-Verlag, Berlin.

——.1988. Paleozoic mass extinctions: shared patterns suggest global cooling as a common cause. *American Journal of Science* 288:334–352.

——. 1989. *Earth and Life Through Time.* W. H. Freeman, San Francisco.

——. 1990a. The general correlation between rate of speciation and rate of extinction: fortuitous causal linkages. In R. M. Ross and W. D. Allmon, eds., *Causes of Evolution: A Paleontological Perspective*, pp. 103–127. University of Chicago Press, Chicago.

——. 1990b. Delayed recovery and the spacing of major extinctions. *Paleobiology* 16:401–414.

Stanley, S. M., and X. Yang. 1994. A double mass extinction at the end of the Paleozoic Era. *Science* 266:1340–1344.

Strauss, D., and P. M. Sadler. 1989. Classical confidence intervals and Bayesian probability estimates for ends of local taxon ranges. *Mathematical Geology* 21:411–427.

Stryer, L. 1975. *Biochemistry.* W. H. Freeman, San Francisco.

Stuckey, C. W. 1978. Milestones in Gulf Coast economic micropaleontology. *Gulf Coast Association of Geological Societies Transactions* 28:621–625.

Sugihara, G., and R. M. May. 1990. Applications of fractals in ecology. *Trends in Ecology and Evolution* 5:79–86.

Sugihara, G., K. Schoenly, and A. Trombla. 1989. Scale invariance in food web properties. *Science* 245:48–52.

Sundquist, E. T. 1985. Geological perspectives on carbon dioxide and the carbon cycle. In E. T. Sundquist and W. S. Broecker, eds., *The Carbon Cycle and Atmospheric CO$_2$: Natural Variations Archean to Present,* pp. 5–59. American Geophysical Union Monograph 32, Washington, D.C.

Sverdrup, H. U., M. W. Johnson, and R. H. Fleming. 1942. *The Oceans: Their Physics, Chemistry, and General Biology.* Prentice-Hall, Englewood Cliffs, N.J.

Swart, P. K., G. F. Healy, R. E. Dodge, P. Kramer, J. H. Hudson, R. B. Halley, and M. B. Robblee. 1996. The stable oxygen and carbon isotopic record from a coral growing in Florida Bay: a 160-year record of climatic and anthropogenic influence. *Palaeogeography, Palaeoclimatology, Palaeoecology* 123:219–237.

Szmant, A. M., and A. Forrester. 1996. Water column and sediment nitrogen and phosphorus distribution patterns in the Florida Keys, U.S.A. *Coral Reefs* 15:21–41.

Tappan, H. 1968. Primary production, isotopes, extinctions and the atmosphere. *Palaeogeography, Palaeoclimatology, Palaeoecology* 4:187–210.

——. 1970. Reply to Phytoplankton abundance and late Paleozoic extinctions. *Palaeogeography, Palaeoclimatology, Palaeoecology* 8:56–66.

——. 1971. Microplankton, ecological succession and evolution. *Proceedings North American Paleontological Convention,* Part H, pp. 1058–1103.

——. 1980. *Paleobiology of Plant Protists.* W. H. Freeman, San Francisco.

——. 1982. Extinction or survival: selectivity and causes of Phanerozoic crises. In L. T. Silver and P. H. Schultz, eds., *Geological Implications of Impacts of Large Asteroids and Comets on the Earth,* pp. 265–276. Geological Society of America, Boulder, Colo.

——. 1986. Phytoplankton: below the salt at the global table. *Journal of Paleontology* 60:545–554.

Tappan, H., and A. R. Loeblich. 1971. Geobiologic implications of fossil phytoplankton evolution and time-space distribution. In R. Kosanke and A. T. Cross, eds., *Symposium on Palynology of the Late Cretaceous and Early Tertiary,* pp. 247–340. Geological Society of America, Boulder, Colo.

——. 1973. Evolution of the oceanic plankton. *Earth Science Reviews* 9:207–240.

——. 1988. Foraminiferal evolution, diversification, and extinction. *Journal of Paleontology* 62:695–714.

Tarbuck, E. J., and F. K. Lutgens. 1985. *Earth Science.* Charles E. Merrill, Columbus, Ohio.

Tardy, Y., N'Kounkou, and J.-L. Probst. 1989. The global water cycle and continental erosion during Phanerozoic time (570 my). *American Journal of Science* 289:455–483.

Thayer, C. W. 1983. Sediment-mediated biological disturbance and the evolution of marine benthos. In M. J. S. Tevesz and P. L. McCall, eds., *Biotic Interactions in Recent and Fossil Benthic Communities,* pp. 480–625. Plenum, New York.

——. 1992. Escalating energy budgets and oligotrophic refugia: winners and drop-outs in the Red Queen's race. In S. Lidgard and P. R. Crane, eds., *Fifth North American Paleontological Convention, Abstracts and Program,* p. 290. Paleontological Society Special Publication No. 6.

Thickpenny, A., and J. K. Leggett. 1987. Stratigraphic distribution and palaeooceanographic significance of European early Paleozoic organic-rich sediments. In J. Brooks and A. J. Fleet, eds., *Marine Petroleum Source Rocks,* pp. 231–247. Geological Society of London, London.

Thomas, C. D. 1994. Extinction, colonization and metapopulations: environmental tracking by rare species. *Conservation Biology* 8:373–378.

Thunell, R. C. 1976. Optimum indices of calcium carbonate dissolution in deep-sea sediments. *Geology* 4:525–528.

Tissot, B. 1979. Effect on prolific petroleum source rocks and major coal deposits caused by sea-level changes. *Nature* 277:462–465.

Tomascik, T., and F. Sander. 1985. Effects of eutrophication on reef-building corals I. Growth rate of the reef-building coral *Montastrea annularis. Marine Biology* 87:143–155.

——. 1987a. Effects of eutrophication on reef-building corals II. Structure of scleractinian coral communities on fringing reefs, Barbados, West Indies. *Marine Biology* 94:53–75.

——. 1987b. Effects of eutrophication on reef-building corals III. Reproduction of the reef-building coral *Porites porites. Marine Biology* 94:77–94.

Tréguer, P., D. M. Nelson, A. J. Van Bennekom, D. J. DeMaster, A. Leynaert, and B. Quéguiner. 1995. The silica balance in the world ocean: a reestimate. *Science* 268:375–379.

Ulanowicz, R. E. 1980. An hypothesis on the development of natural communities. *Journal of Theoretical Biology* 85:223–245.

Umbgrove, J. H. F. 1950. Symphony of the Earth. Martinus Nijhoff, The Hague.

Vail, P. R., R. M. Mitchum, R. G. Todd, J. M. Widmier, S. Thompson, J. B. Sangree, J. N. Bubb, and W. G. Hatelid. 1977. Seismic stratigraphy and global changes of sea level. In C. E. Payton ed., *Seismic Stratigraphy: Applications to Hydrocarbon Exploration,* pp. 49–212. American Association of Petroleum Geologists Memoir 26, Tulsa, Okla.

REFERENCES

Valentine, J. W. 1971. Resource supply and species diversity patterns. *Lethaia* 4:51–61.

——. 1973. *Evolutionary Paleoecology of the Marine Biosphere.* Prentice-Hall, Englewood Cliffs, N.J.

——. 1989. How good was the fossil record? Clues from the Californian Pleistocene. *Paleobiology* 15:83–94.

——. 1995. Why no new phyla after the Cambrian? Genome and ecospace hypotheses revisited. *Palaios* 10:190–194.

Valentine, J. W., and C. L. May. 1996. Hierarchies in biology and paleontology. *Paleobiology* 22: 23–33.

Valentine, J. W., and E. M. Moores. 1970. Plate tectonic regulation of faunal diversity and sea level: a model. *Nature* 228:657–659.

——. 1972. Global tectonics and the fossil record. *Journal of Geology* 80: 167–184.

van Andel, T. H. 1975. Mesozoic/Cenozoic calcite compensation depth and the global distribution of calcareous sediments. *Earth and Planetary Science Letters* 26:187–195.

——. 1994. *New Views on an Old Planet: A History of Global Change.* Cambridge University Press, Cambridge, England.

van Cappellen, P., and E. D. Ingall. 1994. Benthic phosphorus regeneration, net primary production, and ocean anoxia: a model of the coupled marine biogeochemical cycles of carbon and phosphorus. *Paleoceanography* 9:677–692.

Van de Poel, H. M., and W. Schlager. 1994. Variations in Mesozoic–Cenozoic skeletal carbonate mineralogy. *Geologie en Mijnbouw* 73:31–51.

van Straaten, L. M. J. U. 1952. Biogenic textures and the formation of shell beds in the Dutch Wadden Sea. Koninklijke Nederlandse Akademie van Wetenschappen, Proceedings. Series B. *Physical Sciences* 55:500–516.

Van Valen, L. 1976. Energy and evolution. *Evolutionary Theory* 1:179–229.

Vasconcelos, C., J. A. McKenzie, S. Bernasconi, D. Grujic, and A. J. Tien. 1995. Microbial mediation as a possible mechanism for natural dolomite formation at low temperatures. *Nature* 377:220–222.

Vermeij, G. J. 1987. *Evolution and Escalation: An Ecological History of Life.* Princeton University Press, Princeton, N.J.

——. 1989. The origin of skeletons. *Palaios* 4:585–589.

——. 1990. Asteroids and articulates: Is there a causal link? *Lethaia* 23: 431–432.

——. 1994. The evolutionary interaction among species: selection, escalation, and coevolution. *Annual Review of Ecology and Systematics* 25:219–236.

——. 1995. Economics, volcanoes, and Phanerozoic revolutions. *Paleobiology* 21:125–152.

Vidal, G., and A. H. Knoll. 1982. Radiations and extinctions of plankton in the late Proterozoic and early Cambrian. *Nature* 296:57–60.

Vogt, P. R. 1989. Volcanogenic upwelling of anoxic, nutrient-rich water: a possible factor in carbonate-bank/reef demise and benthic faunal extinctions? *Geological Society of America Bulletin* 101:1225–1245.

Volk, T. 1989a. Sensitivity of climate and atmospheric CO_2 to deep-ocean and shallow-ocean carbonate burial. *Nature* 337:637–640.

——. 1989b. Rise of angiosperms as a factor in long-term climatic cooling. *Geology* 17:107–110.

Vrba, E. S. 1984. What is species selection? *Systematic Zoology* 33:318–328.

Walker, K. R., and L. P. Alberstadt. 1975. Ecological succession as an aspect of structure in fossil communities. *Paleobiology* 1:238–257.

Walker, K. R., and W. W. Diehl. 1985. The role of marine cementation in the preservation of Lower Paleozoic assemblages. *Philosophical Transactions of the Royal Society of London* 311:143–153.

Walker, S. E., and J. T. Carlton. 1995. Taphonomic losses become taphonomic gains: an experimental approach using the rocky shore gastropod, *Tegula funebralis. Palaeogeography, Palaeoclimatology, Palaeoecology* 114:197–217.

Wallin, I. E. 1923. The mitochondria problem. *American Naturalist* 57:255–261.

Walter, L. M., and E. A. Burton. 1990. Dissolution of recent platform carbonate sediments in marine pore fluids. *American Journal of Science* 290:601–643.

Wang, K., H. H. J. Geldsetzer, and H. R. Krouse. 1994. Permian–Triassic extinction: organic $\delta^{13}C$ evidence from British Columbia, Canada. *Geology* 22:580–584.

Wehmiller, J. F., L. L. York, and M. L. Bart. 1995. Amino acid racemization geochronology of reworked Quaternary mollusks on U.S. Atlantic coast beaches: implications for chronostratigraphy, taphonomy, and coastal sediment transport. *Marine Geology* 124:303–337.

Werner, D., ed. 1977. *The Biology of Diatoms.* University of California Botanical Monograph 13.

West, O. L. O. 1995. A hypothesis for the origin of fibrillar bodies in planktic foraminifera by bacterial endosymbiosys. *Marine Micropaleontology* 26:131–135.

Wheatcroft, R. A. 1990. Preservation potential of sedimentary event layers. *Geology* 18:843–845.

——. 1992. Experimental tests for particle size-dependent bioturbation in the deep ocean. *Limnology and Oceanography* 37:90–104.

Wheatcroft, R. A., and P. A. Jumars. 1987. Statistical re-analysis for size dependency in deep-sea mixing. *Marine Geology* 77:157–163.

Wheatcroft, R. A., P. A. Jumars, C. R. Smith, and A. R. M. Nowell. 1990. A mechanistic view of the particulate biodiffusion coefficient: step lengths, rest periods and transport directions. *Journal of Marine Research* 48:177–207.

Wicken, J. S. 1980. A thermodynamic theory of evolution. *Journal of Theoretical Biology* 87:9–23.

Wignall, P. B., and A. Hallam. 1992. Anoxia as a cause of the Permian/Triassic extinction: Facies evidence from northern Italy and the western United States. *Palaeogeography, Palaeoclimatology, Palaeoecology* 93:21–46.

REFERENCES

Wignall, P. B., and R. J. Twitchett. 1996. Oceanic anoxia and the end Permian mass extinction. *Science* 272:1155–1158.

Wilde, P., and W. B. N. Berry. 1984. Destabilization of the oceanic density structure and its significance to marine "extinction" events. *Palaeogeography, Palaeoclimatology, Palaeoecology* 48:143–162.

Wilgus, C. K., B. S. Hastings, C. G. Kendall, H. W. Posamentier, C. A. Ross, and J. C. Van Wagoner, eds., 1988. *Sea-Level Changes: An Integrated Approach.* Society of Economic Paleontologists and Mineralogists Special Publication No. 42.

Wilkinson, B. H. 1979. Biomineralization, paleoceanography, and the evolution of calcareous marine organisms. *Geology* 7:524–527.

Wilkinson, B. H., and T. J. Algeo. 1989. Sedimentary carbonate record of calcium-magnesium cycling. *American Journal of Science* 289:1158–1194.

Wilkinson, B. H., and R. K. Given. 1986. Secular variation in abiotic marine carbonates: constraints on Phanerozoic atmospheric carbon dioxide contents and oceanic Mg/Ca ratios. *Journal of Geology* 94:321–333.

Wilson, E. O. 1992. *The Diversity of Life.* Belknap Press, Cambridge, Mass.

Wilson, M. V. H. 1988. Taphonomic processes: information loss and information gain. *Geoscience Canada* 15:131–148.

Wood, R. 1993. Nutrients, predation, and the history of reef-building. *Palaios* 8:526–543.

Woodward, J., and D. Goodstein. 1996. Conduct, misconduct and the structure of science. *American Scientist* 84:479–490.

Worsley, T. R., R. D. Nance, and J. B. Moody. 1986. Tectonic cycles and the history of the Earth's biogeochemical and paleoceanographic record. *Paleoceanography* 1:233–263.

———. 1991. Tectonics, carbon, life, and climate for the last three billion years: a unified system? In S. H. Schneider and P. J. Boston, eds., *Scientists on Gaia,* pp. 200–210. MIT Press, Cambridge, Mass.

Wright, D. H., D. J. Currie, and B. A. Maurer. 1993. Energy supply and patterns of species richness on local and regional scales. In R. E. Ricklefs and D. Schluter, eds., *Species Diversity in Ecological Communities,* pp. 66–74. University of Chicago Press.

Wright, J., H. Schrader, and W. T. Holser. 1987. Paleoredox variations in ancient oceans recorded by rare earth elements in fossil apatite. *Geochimica et Cosmochimica Acta* 51:631–644.

Wyatt, A. R. 1984. Relationship between continental area and elevation. *Nature* 311:370–372.

Zachos, J. C., M. A. Arthur, and W. E. Dean. 1989. Geochemical evidence for suppression of pelagic marine productivity at the Cretaceous/Tertiary boundary. *Nature* 337:61–64.

Zahn, R. 1994. Core correlations. *Nature* 371:289–290.

Zhuravlev, A. Y., and R. A. Wood. 1996. Anoxia as the cause of the mid-Early Cambrian (Botomian) extinction event. *Geology* 24:311–314.

INDEX

Bioturbation; Bivalves; Brach-
iopods
Macroevolution. *See* Evolution
Magnetic reversals, and biostratigra-
phy, 34–35; and mantle plumes
and extinction, 195
Malkowski, K., 199–200
Marshall, C. R., 46, 48–49
McKinney, M. L., 97, 100–101, 110–
112, 190, 207–208
Measurement. *See* Scale
Meldahl, K. H., 46–48, 70
Mesotrophic conditions, 178
Methodology, scientific, xi, 1–7, 39,
213; and biostratigraphy, 29; his-
torical science as narrative, 2–3,
6–7; hypothetico-deductive
method, 4–5; 212–216
Microhabitats. *See* Foraminifera
Microtektites, 76, 79, 191, 192. *See
also* Extinction
Milankovitch cycles, 133, 136–142,
153. *See also* Carbon dioxide; Sea
level
Mitchum, R., 36
Molecular clocks, 2

Newell, N., 192–193
Newton, Isaac, 1, 4, 118–119,
213–214
Nitrogen, 164; isotopes, 153
Nomothetic relationships. *See* Laws
Number systems, 1
Nutrients, 118, 124, 130, 152, 161,
164–165, 169; and alternative
community states, 123–124;
cadmium–calcium ratios, 152; diversification and extinction of
plankton, 160–161, 164–179, 196,
198–201, 202–206; and marine car-
bon:phosphorus (MCP) episodes,
166–167, 169, 174, 177, 179, 181–
183. *See also* Carbon burial; Carbon isotopes; Energy; Extinction;
Phosphorus

Occam's Razor, 1
Ocean circulation, Antarctic Bottom
Water (AABW), 150–152; and ex-
tinction, 194–195, 205–206;
North Atlantic Deep Water
(NADW), 150–153; psychro-
sphere, 135, 149; salinity-driven,
165, 171, 175, 177; surface, 144–
149, 164; thermohaline, 22–23;
135, 147, 150–152, 165–168, 174,
178, 182, 194. *See also* Extinction; Nutrients
Odum, E. P., 114–116
Officer, C. B., 66, 67, 191, 195
Oligotaxy, 145, 172, 207
Oligotrophy, 164, 170, 175, 202. *See
also* Energy; Nutrients
Ontology, 4
Operon model, 125
Oppel, A., 29–32, 81
Origin, of life, 213–214; of phyla,
124–126
Original Horizontality, principle of,
28
Ostracods, and climate change, 129;
and groundwater, 129; and pollution, 129
Oxygen, and diatom blooms, 129;
and microhabitats, 82; and deep
ocean circulation, 166–167, 168,
169, 170, 171–173, 178, 192, 199;
and porewater chemistry, 65, 82,
86, 90–92; redox conditions at the
earth's surface, 155, 158, 166–167,
174, 181, 199; and spallation reactions, 65
Oxygen isotopes, 34, 49, 138–140,
149, 154, 194; Paleozoic glaciation, 165–168

Parameter, 119, 183
Parsimony, principle of, 1
Particle model. *See* Community sta-
bility; Ecological locking
Perched faunas, 203